The Under-Standing of Eclipses

A book inspired by the central eclipse of 1991,
re-inspired by the cross-American eclipse of 2017,
and contemplative of all the needles of shadow
that knit together Sun, Moon, and Earth

Guy Ottewell

ISBN 978-0-934546-74-4

1st printed April 1991. 2nd edition October 1991. Reprinted 1993. 3rd edition 2004. 4th edition 2013. 5th edition 2016.

To Tilly, away for a week

Universal Workshop

www.UniversalWorkshop.com

Others by Guy Ottewell available from Universal Workshop:

The Astronomical Companion
  General guide to astronomy (not annual), with many 3-D diagrams (2nd edition)

Albedo to Zodiac
  Glossary of astronomical terms, with pronunciation, origin, meaning

To Know the Stars
  Young people's introduction to astronomy (2nd edition,#)

The Thousand-Yard Model, or, The Earth as a Peppercorn
  Instructions for a walk making vivid the scale of the solar system

Berenice's Hair
  Story of the stolen tress that became the constellation Coma Berenice

Contents

Cover illustrations

The 2017 August 21 eclipse, pictured on the front cover, is shown with more explanation on page 55.

On the back cover is my rendering of the total eclipse of 1983 June 11. I was lucky to be, with a few others, inside Borobudur in Java, a ruined temple of enormous mass, a pyramidal mountain of sculptured stone representing the levels of existence. Its profile may look in the darkness like a forest of conifers, but those shapes, rising on terrace after terrace, are bell-like stupas in which five hundred calm Buddhas of more than human size sit contemplating the horizon.

Everywhere in Java/ the background is a soaring volcano. Here there are two of them, Merbabu and Merapi, "Tower of Ash" and "Tower of Fire." The day had dawned clear except for a wisp of vapor rising from Merapi; this wisp became a hood of cloud, and all morning it spread toward the Sun. But it did not reach the Sun, and remained a thin frosting, through which, many minutes before totality, burst Venus.

Merapi and Merbabu stand to the east and northeast, so that the false sunset colors of inner eclipse appeared over them; and afterwards the umbra strode away over them. I was again inside the temple before the next dawn, watching as the Sun rose *between* the twin mountains. (With the invisible Moon now preceding it.)

This painting does not, I think, have a very close relation with however the scene literally looked. A dozen people including Indonesian soldiers clustered at the foot of the temple, tall eucalyptus trees divided the sky, and Merapi and Merbabu were not visible from our position on the south side. But, more fundamentally, during the minutes of totality I was aware of being in a different visual world; of trying to memorize colors for which I had no names, which would be as hard to recall or describe as a taste. During the hours of our wait I had begun an outline drawing, which soon rambled to five or six large sheets, so that I had to make myself a diagram of how pieces *a, b, c, d, e* fitted together; and afterwards several people who had been there with me sent me their photographs to help in making a painting. But I didn't do anything about it for eight years, and then those aids proved irrelevant. I remembered that the scene seemed to me to be made of several huge flat pieces stitched together; so I made the picture as a collage of two pieces, anyway. Reproduced here is less than half of the painting's width.

An eclipse, or a comet, may provide the impetus for traveling to a far place, but the other reason is to poke around the country, learn the language. Java and Bali are extraordinarily beautiful lands. I did several hundred sketches, and wrote afterwards a compressed but long account of the eclipse, of Borobudur, and of my wanderings. I have usually arrived at eclipse scenes by means less modern than car or plane, and have been the contemplative (or indolent) kind of observer, taking no photographs and making no scientific studies. Several eclipse chasers have witnessed more than thirty. My own modest record is (not counting partial and lunar eclipses): 1972 July 10 (camping at Cap Chat in the Gaspé peninsula of Quebec); 1979 Feb. 26 (over the snow at Lundar north of Winnipeg); 1980 Feb. 16 (clouded out during totality in the Taita Hills of Kenya); 1983 June 11 (Borobudur); 1984 May 30 (broken-ring annular, at home in Greenville; South Carolina); 1991 July 11 (Sayulita in Mexico); 1994 May 10 (annular, on a road in west Texas); 1995 Oct. 24 (the abandoned city Fatehpur Sikri in India); 1997 March 9 (glimpsed through the fringe of a snowstorm on the road south from Darkhan in Mongolia); 1998 Feb. 19 (on a three-masted schooner in the Guadeloupe Channel south of Antigua); 1999 Aug. 11 (hilltop above the village Shenyurt near Turhal in Turkey); 2002 Dec. 4 (Ceduna on the coast of South Australia); 2003 May 31 (annular, too low and too cloudy in the dawn at St. Margaret's Hope in the Orkney Islands); 2006 March 29 (Mediterranean southeast of Crete); 2015 March 20 (Faroe Islands). .

The frieze along the top of the pages is a movie, or rather two movies: on the left, stages in an annular (ring-shaped) eclipse of the Sun; on the right, stages in a total eclipse. Use your right thumb and let the pages flick from under it. The sky grows darker as the visible area of the Sun's surface shrinks.

The Moon's outline is at first indicated by dashes, but it cannot really be seen until it becomes silhouetted against the Sun. We are looking southward; Sun and Moon move from left (east) to right (west) with the turning of the sky as the day wears on. The Moon moves slightly more slowly in that direction, so that it falls back across the Sun. This is only the apparent motion due to the turning of the Earth; what is really happening in space is that the Moon is moving eastward (leftward) around the Earth.

The movie runs from about 20 minutes before the beginning of partial eclipse to about half an hour after the end of central (annular or total) eclipse. Partial eclipse starts about an hour and a half before the middle of the eclipse. Each frame represents about 3 minutes. But the Sun's apparent motion is really about 15° an hour, or its own half-degree width in 2 minutes, so the band of sky it traverses should be 2½ times as long as we have space for.

The reason why on one occasion the eclipse is annular is that the Moon is farther away from us, therefore appears smaller, and cannot cover the Sun. We show the Moon at its extremes of size, 0.9 as wide as the Sun for the annular eclipse and 1.09 for the total. In many annular eclipses the ring of Sun around the Moon is even narrower; in many total eclipses the excess of the Moon over the Sun is less and totality lasts a shorter time. The longest annular eclipses last longer than the longest total ones; notice that the annular phase spreads over three frames.

The eclipses are as seen from within the track of annularity or of totality, but from a place south of the centerline, so that the Moon's center passes north of the Sun's. Hence in the annular eclipse the Sun-ring is at its widest on the south. In the total eclipse the narrowing crescent slews around toward the south in the last moments, before breaking into an arc of Baily's Beads. The last of these seems, because of your eyes' adaptation to the fading light, to flash out; combined with the first inkling of the encircling corona, this is the effect called the Diamond Ring. The Diamond vanishes, the light-level falls to its lowest, and the full pearly corona seems to flood out from behind the Moon. After five (if you are lucky) precious minutes, the second Diamond—seeming even brighter than the first—bursts with dangerous suddenness through the first valley in the Moon's trailing edge. Light seems to return quicker than it departed; and this brightening sky reclaims the briefly manifested Moon.

Compare the two experiences. Annular eclipse is really a kind of partial eclipse; a fine natural spectacle, but its summit is bland and smooth. Totality is a glimpse into the hall of the gods on the top of Mount Olympus.

The Diamond Ring, seen at the onset of total eclipse, is a popular symbol for "astronomy," along with the star, the comet, the spiral galaxy, the Big Dipper, the Pleiades, and Saturn's rings. Here it is used in the logo of the British Astronomical Association.

Beginning

A total eclipse of the Sun is a sublime experience. It is a moment at which heaven touches Earth and astronomy turns from a solemn polysyllable into an embrace between man and the cosmos. After it, one wants to talk of the otherworldly landscape, the panic of mockingbirds, the explosion of mysterious light, the jubilation of people—not of geometrical groundwork.

Yet an account that does not start at the beginning cannot aspire to be ideal: there would be backward steps and wasted words— "And by the way, we should have explained what the penumbra is . . ."

Hidings and shadings

Eclipses are seen because, in the whirling of bodies around each other which is characteristic of our universe, it often chances that three of the bodies arrive almost exactly in line. Then one of them, as seen from some viewpoint, partly or wholly disappears.

The word *eclipse* is from Greek *ek-leipsis*, an "out-leaving," an absence by a shining body from where it should be seen.* Actually it is applied to two different kinds of disappearance:

(1) The covering or hiding of one body by another. Thus in an eclipse of the Sun the Moon gets in front of it, blocking our sight-line toward it. When the body that becomes hidden (or, rarely, partly hidden) is an apparently small one such as a planet or star, this is called an *occultation* (from the root of Latin *celare*, "conceal").

(2) The throwing of a shadow onto a body. Thus in an eclipse of the Moon what is blocked is not our sight-line to it but the path of light to it; we see it, but with a shadow on it.

Eclipses involving the Sun and Moon (*lunisolar* eclipses), as seen from the Earth, are the kinds that are conspicuous to us. It was for these that the word *eclipse* was invented, and they are what the rest of this book is about. But other bodies can undergo eclipses, though most, being very distant from us, can be observed only with telescopes.

Thus there are transits of planets across the Sun; occultations of stars by the Moon or a planet; rarely, occultations of planets by each other; and apparent dimmings of the light of a star which happen because the star is really an "eclipsing binary"—two stars revolving around each other. All these are eclipses of the hiding kind.

An eclipse of the shading kind is the disappearance of an artificial satellite into the Earth's shadow. If you see a point of light moving slowly overhead within an hour or two after sunset, not blinking in the manner of an aircraft but perhaps varying rhythmically because of its "tumbling" in space, it is probably a satellite, and if on moving a certain distance away from the direction of the Sun it fades, it certainly is—a satellite being eclipsed by the Earth.

The four bright natural satellites (or "moons") of Jupiter can be watched easily in small telescopes as they go through their motions. We see a satellite disappear behind the planet; not quite at the same time, it vanishes into the planet's shadow; later it passes across the face of the planet; and, again not necessarily at the same time, it throws its small shadow on the planet. Thus we see each body (planet and satellite) eclipse the other in each of the two ways (hiding and shading); and four terms for these events have to be used: occultation (of the satellite); eclipse (which here means the shading of the satel-

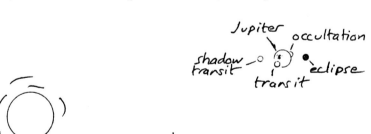

lite); transit (hiding of a small part of the planet); and shadow-transit (shading of a small part of the planet).

This example shows that there are as many as four entities involved in any eclipse: a light-source (such as the Sun); the two eclipsed-or-eclipsing bodies; and the observer, who may or may not be on one of the bodies in the line-up.

However, in our more familiar lunisolar eclipses the locations are reduced to three: in solar eclipses the light-source is the same as the eclipsed body (the Sun), and in lunar eclipses the observer is on the eclipsing body (the Earth).

---

* Other uses of the word *eclipse* in English are metaphorical extensions of the astronomical one. E.g.: "After Ptolemy, Greek astronomy went into eclipse."

Umbra and penumbra

Whenever there is a light-source there is also its shadow: all the space from which rays coming directly from the source are blocked, so that the source or part of it cannot be seen. Actually when we say "shadow" we are thinking sometimes of this volume of space, sometimes of its terminal area, its print on a material surface—that area of the surface that is deprived of light.

A light-source may be point-like, or it may have width. It can hardly be a true point; the stars appear to be points, but only because they are so far away. The Sun certainly is an extended source: it is 1,392,000 kilometers in linear width, and appears about half a degree wide as seen from the Earth.

If a source has width, its shadow consists of two parts: umbra and penumbra.

Two sets of tangent rays define the umbra and penumbra.

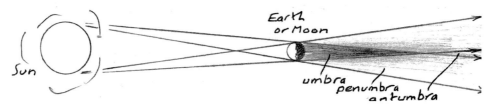

The Latin word *umbra* means "shadow." But we use it for the part of the shadow which is total, that is, from within which no part of the source can be seen.

Penumbra is from the Latin *paene-umbra*, "almost-shadow" (like *peninsula*, "almost-island," and *penultimate*, "almost-last"). But we use it for the partial shadow, which is illuminated by light from part but not all of the source, and from within which part but not all of the source can be seen.

The umbra is the shadow's core, and is uniform, since no light from the source gets into it (unless reflected or refracted from somewhere else). The penumbra is the shadow's outer shell, and grades from almost as dark as the umbra (in inner parts where almost none of the source can be seen) to almost as bright as the non-shadow (in outer parts where almost all of the source can be seen).

When the bodies we are talking about are spheres, the umbra and penumbra are circular in cross-section. The penumbra is always a cone enlarging outward (away from the Sun). As for the umbra, it could also be a cone enlarging outward, less rapidly, if the body intercepting the light (Earth or Moon) were larger than the Sun; or a cylinder, if the intercepting body were the same size as the Sun. But as the Earth and Moon are in fact smaller, their umbrae are cones tapering outward. The farther the Earth or Moon is from the Sun, the longer its umbral cone. It ends at a point. But beyond this point it is continued by an imaginary cone enlarging outward. This is part of the penumbra, but to distinguish it we can call it the anti-umbra, or antumbra.

From any position inside the umbra, the whole Sun is hidden. From each place in the penumbra, part of the Sun is visible—ranging from a speck if you are deep in near the umbra, to the whole Sun with an imperceptible dent in it if you are near the outside.

Each evening, you are briefly in the narrow beginning of the Earth's penumbra while you watch the Sun setting; after the last spark of it has disappeared, you and the landscape around you are in the umbra.*

(Twilight is *not* due to the penumbra. It is the succeeding stage in which no sunlight reaches us directly, but some reaches us indirectly, scattered from high parts of the Earth's atmosphere.)

Huge bodies are not the only ones to have two-part shadows: anything in sunlight casts a penumbra that surrounds its umbra. A shadow that has not traveled far, such as that of your hand held before a wall, is sharp because its penumbra is very narrow. But look at the shadow of a tall pole: it is sharp near the foot of the pole; by the far end it has wide fuzzy borders grading outward—the penumbra. If the pole is tall and narrow enough, its upper part is too small to occult the Sun from any viewpoint on the ground,

so the shadow becomes all penumbra with a core of antumbra. The shadow of a tree-top in winter is a complex of such pen-antumbrae. The ground-shadow of an airplane above a certain height is all penumbra and antumbra.

As for the antumbra, it is like the penumbra in that part of the light-source is visible from it, but what part? From anywhere in the antumbra, rays from the middle of the Sun are blocked, but rays from all around the edge of the Sun reach you past all sides of the Moon. Thus the part of the Sun visible from the antumbra is a ring. It is a ring of even thickness, perfectly symmetrical, only if you are at the central point of the shadow; if you are displaced slightly north of the center, the ring is unsymmetrical, thicker at the north.

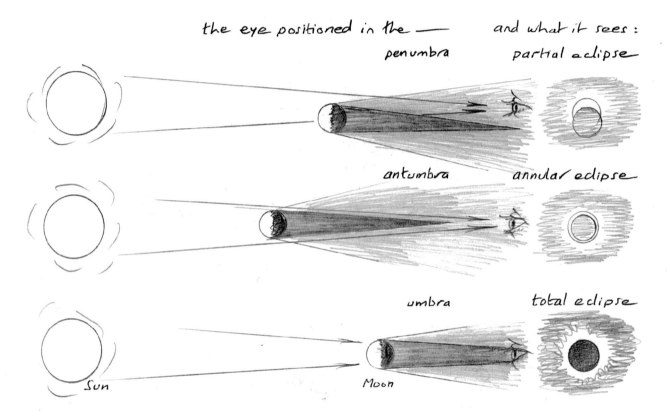

the eye positioned in the —— and what it sees:

penumbra

partial eclipse

antumbra

annular eclipse

umbra

total eclipse

Sun

Moon

Eclipses of the Sun and Moon

Solar eclipses (which is just another way of saying eclipses of the Sun) happen at the time called New Moon, when the Moon passes between the Earth and the Sun. Lunar eclipses happen at Full Moon, when the Moon travels around on the opposite side of the Earth. So these eclipses typically happen in pairs—a solar and a lunar—about two weeks (half a lunar month) apart.

Lunar eclipses are seen at night (or very close to sunset or sunrise), when the Full Moon is in the sky. Solar eclipses are of course seen only by day.

Lunar eclipses can be seen from the whole night side of the Earth: half the Earth simultaneously, then more because the Earth rotates for several hours during the eclipse. But solar eclipses can be seen only from the more restricted geographical area covered by the Moon's moving shadow.

New Moon is normally the time when the Moon is invisible; Full Moon, the time when it is brightest. Part of the excitement of eclipses is paradox: at its brightest time, the Moon goes dark; at its invisible time, it manifests itself out of nowhere as a silhouette.

A lunar eclipse is a throwing of the Earth's shadow onto the Moon, but seen from the Moon it would be a hiding of the Sun by the Earth. A solar eclipse is a hiding of the Sun by the Moon, but seen from the Moon it would be a throwing of the Moon's shadow onto the Earth.

A number used to grade eclipses is their *magnitude*. This answers the question, How *much* of the eclipsed body is eclipsed? and it does so in terms of the body's diameter. Thus in an eclipse of the Sun, the magnitude is the maximum fraction of the Sun's diameter covered by the Moon. It might be, for instance, 0.6 if the eclipse was partial; or 1.001 if the eclipse was total (the Moon spanning all of the Sun's width and .001 more). In an eclipse of the Moon, the magnitude is the maximum fraction of the Moon's diameter covered by the Earth's shadow.

Let's tabulate here the sizes of the three bodies we are talking about and the distances between them:

	kilometers	miles
diameter of Sun	1,392,000	885,000
average Sun-Earth distance	149,600,000	92,956,000
diameter of Earth	12,756	7,926
average Earth-Moon distance	384,400	238,900
diameter of Moon	3,476	2,160

From these we can also calculate the length of the umbrae cast by the Earth and Moon. Clearly they depend on the size of the light-source (the Sun) and illuminated body (Earth or Moon) and the distance between them. If s and e are the diameters or radii of Sun and Earth, d the distance between them, and u the length of the umbra, then $e/u = s/(d+u)$, which can be rearranged as $u = de/(s-e)$.

Thus with the Earth at its average distance from the Sun: umbra length = 149600000 × 12756 / (1392000 − 12756) = about 1,384,000 kilometers. The umbra will be shorter when the Earth in its slightly eccentric orbit is nearer to the Sun (d smaller), around January, and longer when the Earth is farther, around July—by ±23,000 kilometers, which is nearly twice the Earth's width but only $\frac{1}{60}$ of the umbra's length.

The Moon's orbit around the Sun being essentially the same as the Earth's, its average distance is the same, and its smaller diameter yields a proportionately shorter umbra: 149600000 × 3476 / (1392000 − 3476) = about 375,000 kilometers. This umbra too will be shorter around January and longer around July (by ±6,260 kilometers); superimposed on this is a still slighter variation from shorter around New Moon to longer around Full Moon (by about ±1,900 kilometers, half the Moon's width). So the umbra can vary between about 366,000 at a January New Moon and 382,000 at a July Full Moon.

(2) The Earth-Moon system, on a scale of 1 millimeter to 2000 kilometers. Though the Moon is 3/11 as wide as the Earth, most satellites are far smaller in proportion to their planets. Earth and Moon are more like a pair of planets traveling together while orbiting around their common center of gravity or barycenter (which is inside the larger planet, but nearer to its surface than to its center).

(3) The length of umbrae. (*Not* to scale.) Diameters of bodies, and Sun-Earth distance, are given in kilometers.

(1) The Sun-Earth-Moon system, true to scale (at 1 millimeter to 400,000 kilometers), except that the dots for Earth and Moon should be about 10 times smaller! Earth and Moon are shown traveling about 1/12 of their enormous orbit around the Sun. Relative to Earth, the Moon moves in a small orbit counterclockwise from New Moon position around through Full Moon to New Moon; but, relative to the Sun, this change of distance is so insignificant that the Moon's orbit is virtually the same as the Earth's; it never goes backward; it never even makes a concave curve. Moon and Earth proceed together, taking turns to be in the lead or on the outside.

Lunar eclipses

The Earth's umbra is about 1,384,000 kilometers long, the distance to the Moon only about 384,000. So the umbra reaches easily to the Moon and a million kilometers beyond. At the Moon's distance, the umbra is about 2.65 times as wide as the Moon; the penumbra adds about a Moon-width all around. (The exact proportions vary because of the Moon's varying distance.)

The result is that there can be three kinds of lunar eclipse:

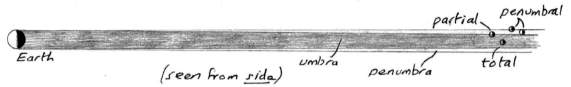

(1) Penumbral: the Moon dips partly or wholly into the outer shadow, the penumbra.
(2) Partial: part of the Moon gets into the umbra.
(3) Total: the whole Moon becomes immersed in the umbra.

Note that these describe the central time-phase of an eclipse. Before the Moon can reach the umbra it must cross the penumbra; so any kind of lunar eclipse begins as penumbral. A purely penumbral eclipse has only one phase. In a partial eclipse there are three phases: penumbral, partial, penumbral. A total eclipse progresses through five phases: penumbral, partial, total, partial, penumbral. The moments when phases begin and end are called "contacts." Thus the second contact in a total eclipse is the moment when the Moon first touches the umbra, so that the partial phase begins.

Both the Earth's shadow and the Moon are moving forward around the Sun, but the Moon is overtaking the shadow: since the Moon goes all around our sky in about a month, but the shadow goes all around in a year, the Moon is moving about 12 times as fast; it moves about its own width in an hour. If the Moon goes centrally through the shadow, it spends more than 6 hours at least partially inside the penumbra, more than 4 hours partially and up to 100 minutes totally inside the umbra. The less centrally it passes, the shorter are these spans.

Now the livelier part: what does a lunar eclipse look like?

The penumbra is so light a shadow that it is hardly a shadow at all. Think how brilliant the Sun is: only a speck of it has to show to turn night into day. Thus it is not surprising that places on the Moon from which most of the Sun is visible look to us undarkened. Only the inner third of the penumbra may be perceptibly gray. Thus penumbral eclipses are hardly even noticed unless and until one edge of the Moon dredges the inner penumbra. Then that edge becomes delicately stained.

The umbra is very different. In the first place, its edge, though soft, is unmistakable, a definite step from light into darkness. And it is curved. When it falls across the middle of the Moon, it shows an arc of as much as 45°, an eighth of the circle. The Greeks

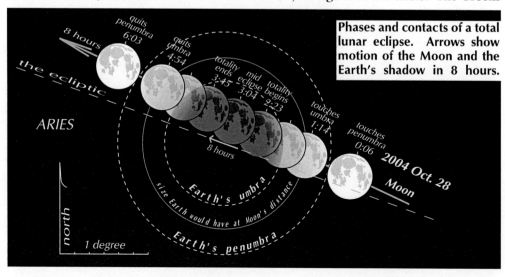

Phases and contacts of a total lunar eclipse. Arrows show motion of the Moon and the Earth's shadow in 8 hours.

realized that this huge shadow out in space (nowhere visible except when it touches the Moon) is the shadow of the Earth; and Aristotle further realized that the shadow's always circular outline is visible proof that the body casting it, the Earth, is a free sphere. Only a sphere would cast a circular shadow whatever the direction of the light; if the Earth were, for example, a disk its shadow would sometimes be a circle, sometimes an ellipse, sometimes a bar.

As for the interior of the umbra, since no sunlight can get directly to it we might expect it to be absolutely black. Instead, nearly black, dark gray, brown, coppery red— it varies between these tones across its area and from eclipse to eclipse. Whence comes this subdued, colored, and unpredictable light?

It is sunlight refracted through the atmosphere of the Earth. (This wonderful insight was first given by Kepler in his *Epitome* of 1618.) If you were standing on the Moon in the Earth's umbra, you would see the Earth totally eclipsing the Sun, but all around the rim of the huge black Earth would be a fine string of ruddy fire: the terminator, the continuous line of sunset and sunrise. The amazing sight has not yet been seen by human eye, but it was photographed in 1967 by the Surveyor III robot lander on the Moon; it was found to have a total brightness 10 to 100 times that of a full Moon as seen from Earth.

The light that has thus skimmed through a band of the Earth's atmosphere is more dimmed and reddened if the band happens to contain more clouds, pollution, or, especially, volcanic dust. On 1883 Aug. 27 Krakatao, in the water between Java and Sumatra, fired five cubic miles of magma into the high atmosphere; more than a year later, on 1884 Oct. 4, the eclipsed Moon was almost black. The eruptions of three Caribbean volcanoes (Pelée, Soufrière, and Santa María) were followed by a lunar eclipse so dark as to be invisible in 1902; that of Mount Katmai in Alaska, by the two dark eclipses of 1913; that of Gunung Agung ("great mountain") on Bali, by those of 1963 Dec. 30 ("the night the full Moon went out") and 1964 June 25; that of El Chichón in Mexico, by another dark eclipse on 1982 Dec. 30.

The Earth, we could say, acts as a circular lens, focusing sunlight onto the Moon; and the Moon acts as a dipstick of the contents of the Earth's shadow, hence of atmospheric conditions. Observers now try to keep standardized reports of how it looks (preferably at the middle of total eclipse), using the scale set up by André Danjon of France:

L = 0: very dark eclipse, Moon almost invisible.
L = 1: dark eclipse, gray or brownish coloration, surface features hard to make out.
L = 2: deep red or rust-colored eclipse, umbra usually has a very dark center and relatively bright outer edge.
L = 3: brick-red eclipse, shadow usually has a light gray or yellow rim.
L = 4: strikingly bright copper-red or orange eclipse, bluish tint where umbra and penumbra meet.

Fractional estimates can be used, such as 1.3; and different parts of the Moon may have different values. People often make reports such as "Probably 2, but the Mare Crisium was invisible and there was an amazingly bright patch along the southern edge . . ."

Solar eclipses

When we consider again the numbers for the sizes and distances of the Sun and Moon, we find one of the most generous coincidences in nature.

The Sun is 400 times wider than the Moon (1,392,000 / 3,476 = 400.5). But it is also about 400 times farther away (149,600,000 / 384,400 = 389.2).

That is why these two disks, the royal pair of our sky, have the same apparent width. Each is about half a degree: half the width of your little finger held at arm's length—much smaller than most people realize, because of the conspicuousness of the Sun and Moon. The Moon can almost exactly fit over the Sun.

This is what grants us eclipses of the most impressive and instructive sort. It is rather

Moon too small or distant— no total eclipse

Moon too large or too near— only fragments of Sun's corona seen and only near beginning and end of eclipse

easy to have a mere eclipse. From the surface of the Moon one could see the Sun disappear behind the apparently much larger Earth—a much-more-than-total eclipse. From the surface of Jupiter one could daily see satellites move across the face of the Sun—a much-less-than-total eclipse. But we happen to be on a planet with many advantages, and one of them is that from it we can see something improbable: only-just-total eclipses, in which the Moon only just blocks the Sun, making possible and prolonging the edge-phenomena.

Put another way, the Moon's umbra reaches roughly as far as the Earth. The Moon likes to tickle us with the point of its shadow. The umbra as we have seen is about 375,000 km long. This is shorter than the Moon's average distance, 384,400 km. However, the distance of the Moon itself varies relatively more widely, ranging from 356,400 to 406,700 km. And these distances are from the center of the Earth, whose surface domes 6,378 km nearer. The result is that the Moon's umbra sometimes ends in space short of the Earth; sometimes reaches right past the Earth; in rarer intermediate cases, it touches the nearest part of the Earth but does not reach the Earth's center.

What is the width of the Moon's spreading penumbra? Since the rays from the distant Sun are nearly parallel, there must where the umbra ends be very slightly more than a Moon-width of penumbra on each side. In other words at this position, which is roughly the position of the Earth, the penumbra must be two Moon-widths or about 7,000 km wide—rather more than half the width of the Earth.

We are now in a position to discern the kinds of solar eclipse:

(1) Partial. The Moon's shadow flies north or south of the Earth, not quite so far as to miss it completely. The penumbra (less than half of it) sweeps one of the Earth's polar regions, and from there the Sun is seen partially eclipsed. But the umbra is aimed into space—so it does not matter whether it reaches as far as the Earth or not.

In the remaining kinds (which we can call umbral eclipses), the shadow passes low enough that the umbra aims at some part of Earth's surface. It may still aim rather wide, making a short track across polar regions; or it may aim near to the center, resulting in a longer track and a better seen eclipse. These eclipses subdivide according to how far the umbra reaches at the time:

(2) Total. The umbra is long enough to reach past the center of the Earth. Where meeting the surface, it is cut off and forms a small round black shadow. From places within this, the Moon is seen completely covering the Sun.

(3) Annular. The umbra does not reach as far as the surface of the Earth. Instead, people on the part of the surface toward which it points find themselves inside the antumbra. This means that they see a ring of the Sun around the Moon. (Latin *annulus* means "ring"; these are sometimes called "ring-eclipses," in German *ringformig*.) The nearer we are to the point where umbra and antumbra meet, the closer the

a partial eclipse

an annular or total eclipse

umbra
penumbra

Earth

Seen from Moon

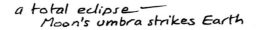

a total eclipse —
Moon's umbra strikes Earth

an annular-total eclipse —
Moon's umbra strikes nearest part of Earth

an annular eclipse —
Moon's umbra stops short of Earth

fit between Moon and Sun. The umbra ends at most 40,000 km above the ground (usually much less), so in actual annular eclipses the ring is very slender. If we could retreat far back in space into the broadening antumbra, the bright ring would become thicker, until we would see the Moon as a mere dot in the middle of the Sun.

Annular eclipse is a transit of Moon across Sun, but because of the Moon's nearness and size it is a short transit: full annularity (with the ring unbroken) can last up to 12 minutes, whereas when Venus or Mercury transit across the Sun they can be completely Sun-surrounded for up to 7 or 8 hours.

More importantly, annular eclipse is a kind of merely partial eclipse, in that the Sun is not fully covered. Daylight is still almost full, the sky does not get very dark, stars do not appear, nor do the vital edge-phenomena. The silhouetted Moon appears like part of the blue sky. Uninstructed people may not even notice that an eclipse is on.

(4) Annular-total, or "hybrid": the uncommon type in which the umbra reaches the Earth's surface, but not as far as the mid-plane. Therefore it cuts only the nearer bulge of the surface. The eclipse is annular for the first part of its course; then the tip of the umbra hits the Earth and the eclipse turns to total for its middle hours; then the umbra loses contact and the last part of the eclipse is again annular. In rare cases ("semi-hybrid") which are close to being all-total, the Moon's distance changes enough during the eclipse that only the beginning of the track, or only the end, is annular. In cases at the other extreme, only the middle moments of the eclipse are total.

Because the Moon's average distance is somewhat greater than the length of its umbra, the umbra more often misses than touches. That is, annular eclipses are slightly commoner than total ones. The print of the antumbra on the Earth's surface can be larger than that of the umbra. One can see an annular eclipse lasting up to about 12 minutes, a total eclipse up to about 7.

The total and annular phases of eclipses are surrounded with partial phases, both in time and space. That is, before the Moon's umbra or antumbra reaches the Earth and after it leaves there are phases, more than an hour long, during which only the penumbra is on. And around the small (county-sized) area where the umbra or antumbra is giving an eclipse, there is a vast (continent-sized) area covered by the penumbra, from within which are seen partial eclipses of infinite variety—their size, orientation, height in the sky, and duration depending on your distance and direction from the eclipse's center.

Partial eclipses yield strange sights (and photographs) low on the horizon (where it can be safe to look at them directly, though not for long). Like the crescent Moon, the crescent Sun may be distorted by refraction and streaked by clouds.

A closer look at classification

Our terms for types of eclipses—which I'll continue in the main to use, since most people do—look as if they are hiding a problem or two:

lunar eclipses	solar eclipses
penumbral	partial
partial	annular
total	annular-total
	total

Several different principles of division are being used. The primary one is as to which component of the eclipsing body's shadow falls on the farther body (at maximum eclipse): only the penumbra, or also the central parts (the umbra or antumbra).

part of shadow	lunar eclipses	solar eclipses
outer:		
penumbra	penumbral	partial
inner:		
antumbra		annular
antumbra and umbra		annular-total
umbra	partial, total	total

Therefore the "partial" solar eclipses can be, and sometimes are, called "penumbral"; annular and annular-total eclipses could be called "antumbral"; and the partial and total lunar eclipses, and total and annular-total solar eclipses, can be called "umbral."

It's partly a matter of aim: if the shadow aims too wide, there is no eclipse; if it aims less wide, the penumbra touches; if it aims less wide again, the umbra or antumbra touches. And it's partly a matter of the Moon's distance: this yields the difference between annular, annular-total and total solar eclipses. The other three boxes are not subdivided this way because it makes no important difference to them: at partial solar eclipses, the umbra misses the Earth and so it does not matter whether it reaches to the Earth's distance or not; at lunar eclipses the Moon's varying distance makes it look larger or smaller, but it is never far enough away to be in the antumbra. (People on the Moon can never see an annular eclipse of the Sun: the Earth is far too large.)

Whereas "penumbral," "antumbral" (if it were used) and "umbral" refer to the shadow and its parts, whatever position in space they are considered from, the terms "partial," "annular," "annular-total" and "total" refer only to the appearance of the eclipsed body as seen from the Earth. Annular eclipses are a subspecies of partial eclipses in that a part of the Sun remains visible. Indeed, for the observer they have more in common with partial than with total eclipses.

We often need a collective term for solar eclipses involving the inner parts of the shadow: the annular, annular-total, and total. The term used is "central." It's difficult to think of anything else (except of course "non-penumbral"); there doesn't seem to be an adjective corresponding to "core"; I found myself toying with "cordal" (from "heart") . . .

One trouble with "central" is that we also use it in a different sense, which has degrees. The eclipse of 1991 July 11 was "almost perfectly central"; others are "less central" until they become "marginal." This kind of centrality relates to the exact center of the shadow, that is, its axis. When that is what is meant, it is better to refer to "axial" eclipses. For eclipses that are central in the sense of involving the umbra or antumbra do not necessarily involve the axis.

In total lunar eclipses, the umbra is nearly three times as wide as the Moon, and so the Moon can move completely inside it with or without touching the shadow's very center. So these eclipses could be divided into axial and non-axial total eclipses. We don't bother to do so, because the axial eclipses appear as merely the longest of the total ones; they don't differ in kind, as total does from partial.

But in solar eclipses, the umbra or antumbra is a very much smaller patch around the axis. So at first it doesn't occur to us that there is any difference between umbral (or antumbral) eclipses and axial ones: if

total lunar eclipses —
umbral but
not axial

umbral
and axial

the umbra touches the Earth, the axis does. This is true except for rare borderline cases.

The axis of the shadow can graze so close outside the northern or southern edge of the Earth's globe that, though it misses, part of the umbra or antumbra (less than half of it) does touch. Therefore from some icy location a total or annular eclipse could be seen (very briefly, and very low to the horizon), even though the eclipse is not axial. The track of totality consists of a short semicircular patch: it has no northern edge (or else no southern one). Such an eclipse hovers on the border between total (or annular) and partial.

umbral but not axial

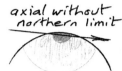

axial without northern limit

Another variation—and it, too, happens—is that the axis does touch, but not quite the whole of the umbra.

And finally, it is just possible that one of these variations affects an eclipse which is transitional in the other sense: that it is annular-total. In other words, the tip of the umbra not only would graze the circular profile of the Earth, but would come very close to reaching just that far: so that the eclipse would start annular-but-not-axial; become for a middle moment total-but-not-axial; and end again annular-but-not-axial; all over a very short semicircular track, from which the exquisitely thin ring (or the upper half of it), then the brief flare of totality, then the ring again, would be seen right down on the horizon. Jean Meeus wrote a paper (in Dutch) to show that this can happen—once in 250 million years.

Partial lunar eclipses—which of course are not at all uncommon—resemble in structure (though not in size) the very rare umbral-but-not-axial solar eclipses: the inner shadow touches, but its center does not.

With eclipses, as with many other things, some of the most intriguing cases lie at the transitions between or the overlap of our categories: between shadow cones that just miss the globe and those that just graze it, cone points that barely reach and barely fail to reach, and Moon outlines that barely cover and barely reveal the Sun's edge.

Borderline cases are sensitive to slight differences in calculating methods or in numbers assumed for such things as the diameter of the Moon. That is why there is occasional disagreement as to which class an eclipse falls into, or indeed whether it was an eclipse at all.

The occurrence of eclipses—eclipse seasons

The Earth moves around the Sun in a plane. The plane has a well-known name—which for a moment we'll hold back, calling it just the Earth's orbital plane.

Meanwhile the Moon orbits around the Earth. If it moved around in this same plane, then at every time of New Moon the Moon would block our view to the Sun, and at every Full Moon our shadow would fall on it. That is to say, eclipses would happen at every New and Full Moon—12 or 13 of each kind a year. They do not, though they happen more often than people tend to think: at each sixth (or sometimes fifth) New and Full Moon.

The reason is that the Moon is traveling in its own orbital plane, which has an inclination, or slope, of about 5 degrees to the Earth's. (To be more exact, the inclination varies between 4.98° and 5.3°, over a period of about 173 days.) As the Earth proceeds in its vastly longer orbit around the Sun, the Moon buzzes around and around it on this slightly sloping orbit, as if the Earth were wearing a belt set on a slant.

The two opposite places in the Moon's orbit where it crosses the Earth's orbital plane are called its ascending and descending nodes. (Here "ascending" means "moving to the north," which isn't "up" for everybody; but impartial terminology would be difficult to find.)

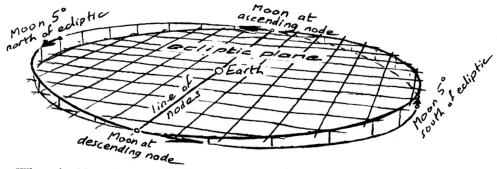

When the Moon is at a node, its center lies exactly in the plane. For about 5 hours after it has passed the node, some part of the Moon's body is still cut by the plane. And for about a day after, some part of the Moon is still as near to the plane as some part of the Earth is. For the rest of its two-week journey from node to node, the whole Moon is well north or south of the plane. Half way between the nodes (at the points of greatest northernmost and southernmost latitude) the Moon is out of the plane by as much as 10 times its own width.

There can be an eclipse only when the Moon is in or near the plane. And that is why this plane, which is fundamentally the Earth's, came to be called the *ecliptic*.

The ecliptic has significance for far more than just eclipses: it is essentially the plane in which we see the whole solar system circulating: the Sun exactly, the Moon and the planets approximately, and, less strictly, the asteroids and comets. But, of all this activity, the eclipses caused by our great satellite's crossing of the plane loomed most impressive to earlier astronomers.

There can be an eclipse, then, only if the Moon happens to arrive at or near one of the two nodes when it arrives at its New or Full position.

The hinge-like imaginary line across the orbit, connecting the ascending and descending nodes, is called the *line of nodes*. As the Earth-Moon system travels around the Sun, this line remains approximately fixed in space. That is, though moving around the Sun, it remains fixed in direction, rather like the side-rod of a steam locomotive, instead of revolving like a spoke of a wheel. Therefore there are just two times in the year, about six months apart, when the line of nodes points to the Sun. These are the *eclipse seasons*: the only times when Earth, Moon, and Sun can be lined up not only in two dimensions but in the third, so that there can be eclipses.

Let us take the liberty of naming these two eclipse seasons "Ad" and "Da." In season Ad, the orbit is positioned so that the ascending node is on the Sunward side of the Earth, the descending node on the outer side. Thus it is when the Moon arrives at its ascending node—climbing northward—that it eclipses the Sun; then as it comes *line of nodes not aligned with Sun — no eclipse possible*

Earth's orbit

First eclipse season (Da) — line of nodes aligned with Sun

Moon's orbit

ascending node

descending node

Sun

second eclipse season (Ad) — line of nodes again aligned with Sun

around to the descending node it is itself eclipsed. At a Da season, the solar eclipse happens at the descending node, the lunar eclipse at the ascending node.

Because the Earth and Moon have size, the line-ups do not have to be exact. So an eclipse season has length (about 38 days). And thus more than one eclipse occurs before the line of nodes moves too far away from the Sun. Usually there is one eclipse of each kind (solar and lunar, in either order).

But in about 1 in 7 eclipse seasons, one of the eclipses is so near the midpoint of the season (the exact time when the nodes are in line with the Sun) that it is flanked by two eclipses which almost missed, making 3 in all (one solar and two lunar, or vice versa).

The line of nodes does not really remain fixed in space. Like many features of the Moon's orbit it is subject to continuous change. It gradually twists, or "precesses," backward (that is, westward: clockwise as seen from the north), the opposite direction to the Moon's own motion. Therefore the next time this line becomes aligned with the Sun—the next eclipse season—is reached not in six months but slightly sooner, by 9.3 days.*

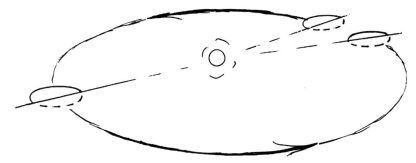

(Modern planetarium projectors include this "regression of the nodes" in their mechanism for the Moon, and are able to show it very clearly by speeding the Moon's motion.)

Thus the length of this span (from the center of one eclipse season to the center of the next) is 173.3 days. We can call it the *eclipse half-year*. It is an important rhythm in the Moon's orbit; we already mentioned it as the period over which the orbit's inclination varies. The earliest calendars in Mesopotamia had years of six lunar months, perhaps based on noticing that eclipses of the Moon came six Full Moons apart. Later, as in the Jewish calendar, the two years, one beginning in autumn with the month Tishrei ("beginning") and the other in spring with Nisan, were joined; but a trace of the old six-month calendar is that the civil year starts with Tishrei, the ecclesiastical with Nisan.

According to different authorities, the world was created on 3761 B.C. Nisan 1 or on 3760 Tishrei 1.

Because this span between eclipse seasons is less than half of an ordinary (solar) year, these seasons move backward through the year. They move back by 18.6 days (twice 9.3, as there are two eclipse seasons per year). If it were not for this, eclipses would always happen at the same two times of year. We would get eclipses in, say, April and October, year after year, and never any others. But, because of the shift, we can get eclipses on any day. For example, in 2017 an eclipse season falls in February, in 2018 it straddles January-February, and by 2019 it is entirely back into January. The same shift pulls the other eclipse season from August to July-August and then July.

There can come a time, as in 2019-2020, when an eclipse season straddles December-January. Then, as in 2020, the next season, being less than 6 months ahead, is in June-July, and a third, or part of it, can fit into the following December.

Having called the span between successive eclipse seasons an eclipse half-year, we can call two of them an eclipse year. It is the interval between two eclipse seasons of the same type (Ad or Da). The eclipse year is 18.6 days shorter than the solar year. Thus 19 eclipse years fit into a little more than 18 years—which is another way of stating the rate at which the eclipse seasons keep shifting earlier in successive solar years.

So here is the combined result of all this, in numbers of eclipses:

In most years (about 70% of them) there are just 2 eclipse seasons, each containing just 2 eclipses, making 4 in all. This is your common foursquare four-eclipse year. There are never fewer eclipses in a year than 4.

But, as we have seen, occasional eclipse seasons have 3 eclipses, and occasional years have 3 eclipse seasons or parts of them. Thus in 2016 the second of the two eclipse seasons had 3 eclipses, so the total was 5. In 2019 the other effect applies: the first eclipse of a coming eclipse season squeezes in at the end of the year, and it is for that reason that there is a total of 5 eclipses. In 2020 both effects apply: one eclipse of the previous year's last season squeezes into January, and then in the middle of the year there is a three-eclipse season, so the total is 6 eclipses.

Finally, there is the rarest kind of year, such as 1973, 1982, and not again until 2038: there is a three-eclipse season, and it is centered close enough to the middle day of the year (July 2) that a whole two-eclipse season fits into January and another into December: result, 7 eclipses. This is the maximum.

(In some discussions, lunar eclipses of the slightest kind, the penumbral, are not counted as eclipses at all. On that basis, many eclipse seasons have only one eclipse—a solar with no lunar—and the number of lunar eclipses in a year can fall to 1 or 0. For instance in 2020 all four lunar eclipses are penumbral. This makes the number of lunar eclipses much lower than that of solar ones—less than 2/3. I do not adopt this reckoning, because penumbral lunar eclipses, though barely observable, are geometrically the counterpart of partial solar eclipses. Omitting them spoils the symmetries.)

If all years were of the common 4-eclipse kind there would be 400 eclipses in a century. Because of the 5-, 6-, and 7-eclipse years, the total number of eclipses per century is about 454.

One might think that, with 4½ eclipses happening a year, we are in the midst of a veritable lantern-show—the Moon exhibitionistically popping from white to black as it skips around us! This is not reality. Many of these eclipses are of slight interest. Many are visible only from inaccessible parts of our planet. And the most interesting—or rather the intense central phases of one kind—are visible only for very brief times in restricted and often remote regions.

Penumbral lunar eclipses, except the deepest, are not appreciable even if you know about them and look hard at them. Partial and total lunar eclipses are very interesting, but are perhaps looked at less than they deserve because they take place during the night. All lunar eclipses are observable over a whole hemisphere of the planet (really more, since the planet turns during the eclipse), so we do not travel specially to see them.

Solar eclipses that remain only partial touch mostly the inaccessible polar regions such as Greenland, Siberia, Antarctica, and the Arctic and Antarctic Oceans. The partial phases of other solar eclipses are visible over wide areas—up to more than half of the Sun-facing hemisphere. But unless and until a partial eclipse becomes deep, most people will not notice any darkening of the sunlight, and will not look attentively to detect the dent made by the Moon in the outline of the Sun. Of the wide areas over which such eclipses are visible, the outer fringes have only a slight eclipse. If one is going to travel at all, it would not be worth going anywhere but to the restricted central band where the eclipse becomes more than partial.

Then there are the annular eclipses. They are spectacular—burning rings in the sky—and are surprisingly frequent. Yet few people are aware of them and fewer take the trouble to go to the narrow areas where they will be seen as actual rings. For, as with partial eclipses, they fail to quench the Sun. They happen in a daytime sky which remains a daytime sky.

It is the *turning of day into night* that makes the crucial difference. The eclipses that do this are in a league by themselves.

Some safe ways of watching the Sun. NOT safe are sunglasses, colored or soot-b;ackened glass, exposed film, filters, stopped-down telescopes, etc. Looking at the Sun is dangerous to the eyes at all times, but we have an instinct not to; the danger during eclipses arises because we are tempted to look longer, and the dimmer light gives a false sense of safety. During totality there is no danger.

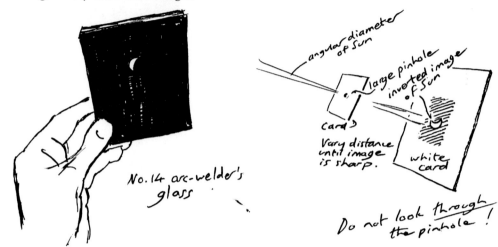

No.14 arc-welder's glass

angular diameter of Sun

large pinhole

card

Vary distance until image is sharp.

inverted image of Sun

white card

Do not look through the pinhole!

Phenomena of a solar eclipse: the partial phases

Enthusiasts and scientists travel to places like Siberia, New Guinea, or the Sahara in order to be within the Moon's core shadow for times from seven minutes down to a few seconds. We will try to show why.

Waiting in the chosen spot and looking at the Sun by means of a pinhole or other safe device, you keep checking for the first imperfection in the brilliant outline. The Moon, falling back across the Sun, is moving, relative to it, from west to east. So the first dent will be in the Sun's west side. What does this mean in terms of right-and-left? If you are in the northern hemisphere and looking roughly south-ward, the Moon's motion west-to-east is right-to-left; the dent will appear on the right. (Vice versa for looking northward from the southern hemisphere: the same motion appears left-to-right, the first dent will be on the left.)

If the time is morning, the Sun with the Moon ahead of it is slanting up the sky. The dent will appear above the middle. Vice versa for the afternoon.

When you are first sure you see this dent, the pulse cannot help quickening. Despite your modernity, you are awed that the prediction is coming true to the minute. Out of the midst of a sky containing nothing but the Sun and its glare, the ponderous presence of the Moon stealthily makes itself felt. Suddenly you know it is really there, hanging at some cavernous distance out in space, like a great bead on a rod running up from you to the Sun. There is a temptation to suspect that the stir of emotion is from some phys-ical stress on venturing into the line of a mighty conjunction—Sun, Moon, heart, Earth!

But you do not really see the Moon. A common conception, appearing in comic-book scenes of eclipses, is that a huge black disk pops into the sky alongside the Sun and moves rapidly to blot it out. This is certainly not so. To a quick glance, the core of the sky is still what it was: a widespread furnace of light, within which it takes optical devices to find the circle of the Sun, and a careful look to find the small dent. What seems to be encroaching on the Sun is merely a part of the dazzling sky around it. The atmosphere around the shaft of direct sunlight is still full of scattered sunlight. Not till almost the center of the eclipse will the circle occupied by the Moon become black.

The next stage, the partial phase of the eclipse, is much the longest. It takes about an hour, if you are near the beginning or end of the eclipse track, or an hour and a half if you are near the middle. During all this time the bite pushes deeper.

Look at the ground under trees. Light passing through a small opening, such as a pin-hole, takes the shape not of the opening but of the light-source. Usually, therefore, all the smaller and sharper light-patches on the ground are elliptical or, if the Sun is over-head, circular—not the polygonal shapes of the openings. Where the openings are higher, the ellipses are larger and fuzzier and overlapping. If a cloud moves over the Sun, the cloud's shadow moves over every image simultaneously—in the opposite direction. We are familiar with this dappled light and may not consciously notice it. But now as the Sun changes its shape so do the light-patches: they become crescents. Half an hour into the eclipse this grows startling: people grab each other's arms and point down at the myriad pinhole pictures.

It's not only gaps in shadow that become crescent-shaped: so do small shadows themselves. The shadow of a bee or an airplane at some height is normally circular; during partial eclipse it is a crescent. The shadow of your fingertip becomes a claw.

Sun-images on a tuliptree leaf through holes in other leaves.

The air cools as the Sun is shut off. Dew may form. Clouds, unfortunately, may form. An "eclipse breeze" may spring up as the land grows cooler than neighboring water. In other circumstances, all winds may drop to a calm, like the eye of a hurricane.

Daylight dims—but when would you first notice this if you didn't know an eclipse was happening? It not only dims, it becomes strange; it is like evening but not. We associate evening light with long shadows; but the short shadows of the high Sun not only have not lengthened, they have stealthily vanished. To me it seems that the reduced light concentrates itself in oily sparkles on leaves, telescopes, and people's faces.

And what color is this ambient light? Some say dusky, some bluish, some greenish; perhaps it is olive. If the Moon, like the Earth, had an atmosphere, the narrow arc of light grazing through it would be reddened. The Moon has no atmosphere, so such light as comes to us comes in an uninterrupted line from the Sun and is the Sun's usual yellow-white—or perhaps not quite. Not only does the outer part of the Sun's image appear slightly less bright because angled away from us, but the surface layers are less dense and hot than those below them; now the light coming to us is only from those outer layers and has a different mixture of wavelengths.

As the light-level drops, automatic street lights come on. Lights that were already shining become more prominent and may be noticed for the first time. On the shore of the St. Lawrence (1972), only far into the eclipse did I notice camp fires scattered among the rocks: people lit them as they became cold, or I had not seen them in the sunlight.

Cattle walk home; birds roost; evening flowers open; bats and night-flying moths come forth. These reactions in the natural world are famous and the list gets longer; I have seen some of them (moths, cattle) but I suspect the counterexamples don't get mentioned. Plants and animals usually don't take a thunderstorm for night; and they tend to be governed by internal clocks that are based on the length of daylight and can't be reset in a matter of minutes.

But more conscious beings certainly react. Dogs bark; people may lock themselves in rooms, or beat gongs, or sound sirens, or applaud.

Or, of course, the sky may be agonizingly clouded. But this is the gamblers' risk that eclipse-chasers take. Some of the most thrilling eclipses have been those where a rift opened at the last moment.

Venus may become visible, as a pinprick in the deepening blue, ten or even as much as thirty minutes before totality; sometimes even through thin cloud. (It is a thrilling moment when the cry of "Venus!" goes up; I remember, in 1983, three Indonesians hugging each other, their upturned faces glistening with, it seemed, Venus's light.) Other stars and planets may appear later, depending on what part of the sky is above the horizon; but only Venus and much fainter Mercury are always near the Sun (Venus up to 47° and Mercury up to 28° east or west of it) and so only they are always in the same hemisphere as the eclipse. Jupiter and some first-magnitude stars have been found by various techniques in daytime skies, but in general only Venus—always the brightest planet and far brighter than any star—appears outside totality.

The brink of totality

In the last minutes things, happen quicker and quicker. The crescent of the Sun dwindles finer. From a semicircle, it shortens in length *faster* as the end approaches. The visible surface of the Sun is called the photosphere ("light-sphere"). A graph of the area of it still in view would, during a partial eclipse, make an inverted bell-shaped curve, but during a total eclipse it does not: it falls more and more steeply, until it bottoms out at zero. This explains why the light-level in the sky drops with increasing rapidity.

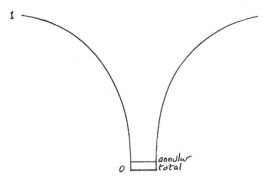

Graph of the fraction of the area of the Sun's disk visible during a central eclipse.

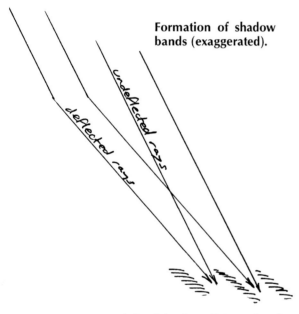

Formation of shadow bands (exaggerated).

Shadow bands are elusive ripples of alternate light and dark, flitting over the ground or the sides of buildings. They are probably always present, but seen only by some eclipse-goers. There have been some fine drawings of them, but they have proved difficult to photograph, because of their movement and their low contrast in the general low light. I first saw them on the snowy plain of Manitoba in February 1979, and not again until I borrowed the idea of Philip and Phylis Morrison. They used to take to eclipses their "Tycho Memorial Instrument" (so named because Tycho observed before the invention of the telescope) and it was always "the instrument of largest aperture present": a bedsheet—queen-sized, of course. Spread it under trees during the partial phases, where it clearly shows the "half-closed eye" shjapes of the small crescent images; then bring it into the open to watch for the shadow bands several minutes before totality.

Vague and jumbled at first, they grow more organized, and flourish in the last twenty seconds. They can give a feeling that the whole scene is under water; indeed, they are akin to the trembling net of light on the bottom of a sunlit pool. They move only at running or walking speed; how can they be related to the great Shadow passing over at a thousand miles an hour? They are not really shadows of anything. A clue is that they are aligned parallel with the sliver that is left of the Sun. It is as the sliver gets narrower and straighter just before totality that the bands become more regular, definite, and close together (from many feet down to about four inches). Stars twinkle because of the turbulence in the atmosphere: their rays are slightly deflected by moving cells of different density in the air. The light from other sources is subject to the same deflections, but when the source is wide and bright they take place within one image and are undetectable. With the Sun reduced to a narrow string—like a string of stars—the deflected rays reach the ground as an addition of twinklings: an interference-pattern. Its movement is that of a wind somewhere above us; it may be diagonal to the bands, but we cannot perceive any motion along the bands and therefore see only the sideways component.

Colors of sunset: you become aware of them on the horizon, but when and where they started is harder to say. The inner penumbra is increasingly dark; the Sun is setting behind the Moon. Normally, the light that comes to you straight from the Sun is white; the light that has taken somewhat indirect routes, scattering off particles in the atmosphere, is predominantly blue because more of blue than of longer wavelengths scatters sideways. At sunset the light coming frontally from the Sun (slightly curved by refraction) has taken a long path through the atmosphere so that most of the blue has been scattered out of it: what is left is dominated by the red, so that the mixture appears gold, orange, or red. When you are in the inner penumbra or the umbra, the atmosphere above you has been almost cut off from its light-supply: almost no light is reaching it directly (white) and little more by nearly direct routes (blue), so that the sky is a blue verging on black. But almost the usual smaller quantity still reaches you horizontally, from regions farther out in the penumbra where the sunlight is still bright. This light, turned golden in the usual way by long travel through the low atmosphere, is always present but is normally drowned by or combined with the blue light from higher to produce pale white; now it is exposed. Thus though the Sun is setting behind a small horizon high in the sky, the sunset colors appear along the usual low horizon.

The umbra itself may be seen arriving. The vast shaft out in space meets nothing to make it visible, though its shape is worth being conscious of: if visible it would seem to taper drastically from you to the Moon, but in reality it tapers from the Moon to a relative needle-point near you. The round foot that intersects the atmosphere manifests itself, especially if there are clouds or haze for it to stain, like a cylindrical storm a hundred miles wide. You have a chance of seeing it if you are on high ground with lower country or sea to the west. At Borobudur in 1983 there were two ridges to our west, and I thought I might see the jagged outline of the farther ridge darken moments before the nearer; but if it happened it was while I blinked. The umbra is a storm traveling at more than a thousand miles an hour: as soon as near enough to see, it will be on you in a rush. It should seem to rush upward, like a giant striding over you.

The corona, which will come to dominate the view during totality, has been glimpsed by keen observers as much as 50 seconds before. It is a faint white light outside the Sun's circle, growing fainter as it spreads outward. There are reports that, in very clear air, usually from high sites, the corona can already be seen spreading past the Moon; in other words the Moon is dimly silhouetted against it, even before totality. In that case this is the stage at which the whole circle of the Moon becomes traceable.

Baily's Beads. The dwindling of the Sun to an ever finer crescent, which goes on all this time, is almost excruciating to watch, but its death does not come by smooth dwindling. It breaks, first near its "horns" or tapered ends, like a thread of liquid evaporating. Then it breaks in as many as a dozen places, into brilliant fragments. Francis Baily carefully observed this phenomenon in the annular eclipse of 1836, and studied earlier descriptions of it. It happens because the Moon is encrusted with mountains, which reach upward through the bright crescent and split it. Sunlight comes lancing only through the valleys; each Bead represents a cove or ravine or mountain pass in the lunar wilderness. An additional optical effect exaggerates the peaks: each seems to be drawn outward to meet the edge of the Sun, as if that were a liquid with surface tension. Something similar is seen in the "black drop" appearance of a transit of Mercury or Venus, as it reaches the Sun's edge and seems to be sucked toward it.

(The Moon is farther than the Earth or Sun from being a smooth spheroid. The mountains, measured from the floors of the Moon's "seas," range up to about the same absolute heights as the Earth's, but are larger relative to the Moon's smaller size, and sharper, because less eroded or pulled down by gravity. The Earth is said to be relatively smoother than a billiard-ball. The Sun has enormous flares and holes but its size and distance make it smoother still.)

The sky steeply darkens to a passionate ultramarine. About this time the Moon at last attains blackness. Can we even now say that it has become visible?—perhaps not, for though we are looking at its Earthward surface, we see nothing.

Light seems sucked in behind this black monster, this absence, leaving only the faint rim of the corona, and the arc of beads.

The beads persist for a few seconds, some shifting and changing, some winking out; then rather suddenly they all die. Except one. This last point of the Sun left uncovered at the Moon's advancing edge may survive only a second or less beyond the others, and must be smaller and fainter than those that have disappeared, yet because of increasing contrast with the blackening background, and our eyes' adapting to lower light, it seems larger and brighter than them all put together: it flashes out to half the size of the Sun. This—the combination of the dim but growing coronal circle and the one dying flash of photospheric brilliance—is the gorgeous effect called the **Diamond Ring**.

Totality!

The Diamond is extinguished: in that instant, totality begins.

Because the light-level has dropped so suddenly, it seems like night— "Darkness at noon." A Actually it is brighter than full Moonlight; like twilight about half an hour after sunset.

At the very borderline of totality there is something to catch: you may fail, but this is the only chance (except for the corresponding moment at the end of totality). It is the **chromosphere** ("color-sphere"), the Sun's inner atmosphere. It is a thin shell (2000 to 3000 kilometers, less than half the width of the Earth). The Moon finishes covering the photosphere; with the dazzle gone, the chromosphere is visible as a red streak; seconds later, the Moon covers it too. To analyse the light of the chromosphere and thus study its nature, astronomers had to grab what was called the "flash spectrum," by slipping a prism in front of a telescope's lens an instant before totality. They found that the red color comes from the strong emission-line of hydrogen-alpha.

The central spectacle

As your eyes adjust, new light seems to come welling out from behind the Moon. This is the Sun's outer atmosphere, the **corona**. Even its inner part is 500,000 times fainter than the photosphere. Invisible unless the photosphere is masked out, it and the chromosphere had to be discovered at eclipses. The corona, being farther out and much larger, was known centuries earlier, but until the nineteenth century and photography there was argument about whether it was attached to the Sun, or the Moon, or was in our own atmosphere, or was an optical illusion. Its name ("crown") was given by Giovanni Domenico Cassini, describing the eclipse of 1706 May 12. The word used might have been "halo," for to some it glorifies the Moon like the head of a saint.

The bursting of the corona into broad view is indeed the crown of an eclipse, the climax of euphoria among the human watchers. It stands out like the manifestation of the specialness, the discoveryfulness of eclipses. And it stands there throughout the minutes of totality, apparently a great splash of light, certainly now the brightest left.

The corona is a pearly or milky white. It is a thin yet extraordinarily hot gas. That is, its temperature is in the millions of degrees—many times hotter than the photosphere—but this only means the gas particles are moving very fast: there are so few of them per unit of space that the heat content is low and they would feel only warm. The corona can be traced outward for several Sun-radii, sliding down a gradient of brightness. Actually it extends millions of miles more, merely becoming thinner; in fact it becomes the solar wind, streaming in all directions from the Sun, so that the Earth is inside it. Near the Sun, the corona is structured into denser curving filaments or streamers. The larger gaps between them are "coronal holes," usually over the Sun's polar regions. Photographs at a single eclipse can make the corona look quite different in size and shape, giving the impression that it must have changed from minute to minute; but this is the effect of different photographic exposures. The structure does vary greatly from one eclipse to another: like many other things it is related to the roughly 11-year cycle of sunspots. Near times of sunspot maximum, the corona tends to be stout and round; near times when sunspots dwindle in numbers (and shrink toward the Sun's equator) it collects itself into a few broad "petals," with shorter brushes over the Sun's poles.

Binoculars are very dangerous outside totality; keep them firmly shut away. But during totality they are the best way to enjoy the details of the corona.

Prominences may stand in view like pink tongues or flames. Count them and note their positions ("ten o'clock" from the top . . .). If five show around the Sun's limb, there may be twenty at the time on its sphere. As with the corona, early observers could hardly believe that these "red mountains" are on the Sun, since at such a distance they would have to be horribly large. But they are: they rise 300,000 kilometers or more, arches and fountains and "anteaters" to which the Earth would be an ant, a hundred times smaller. Because they are so tall they can appear on all sides of the Moon at once. Seeming to leap out of the chromosphere and pour back, they really hang in the corona, formed by magnetic differences near sunspots. Like the chromosphere they have the rosy tinge of hydrogen, but from another emission-line in them a previously unknown element was inferred by Norman Lockyer at an eclipse in 1868 and named helium (from *helios*, the Sun). It was not isolated on Earth until 1895, and is the second simplest and second most abundant element in the universe. (Another element guessed at in this way, "coronium," did not exist.)

Sky and landscape around

So much for the serious work, the features you really ought to look for! But if the eclipse is longer than half a minute you can take some time away from the focal spectacle and glance around at the rest of the scene, on Earth and in the sky: the 360-degree sunset, the transformed landscape. You are in a place outside usual experience, a place you may never be in again; as with a dream, you may perceive now the aspects of its strangeness but may carry out no means of remembering or describing them. They must pertain to tones and colors. Some say that objects are flat, tomb-like, and colors greenish. On walking indoors from bright sunshine I see a green haze, perhaps because it is complementary to the magenta that came through my eyelids outside; and passing from sunshine into the umbra is like entering a room.

The first sight of an eclipse in the twentieth century, early in the morning of May 18, 1901, on the island of Mauritius, was accompanied by a rainbow. It was an unearthly rainbow containing a bright pink line—a spectrum from the Sun's atmosphere.

Look around for stars and planets. Only the few brightest are likely to become visible, so to guide your eyes to them, and to know which they are, you need the help of a sky chart for the time of the eclipse. Mercury, like far brighter Venus, may appear because it is near the Sun; the other bright planets, Jupiter, Saturn, and Mars, may or may not be in the above-horizon half of the sky. If the eclipse is around July you may see the normal "winter" stars Sirius, Procyon, Rigel, Aldebaran; if around January, a "summer" star such as Antares; if in the northern hemisphere, perhaps Capella, Arcturus, Vega; if in the south, Canopus and Alpha Centauri. These are all first-magnitude stars, but people who took the precaution of wearing sunglasses during partial eclipse have been able during totality to see stars as faint as fourth magnitude (of which there are several hundred!). Seeing stars in daytime—stars we are used to seeing on the other side of the year—is more fun than it is important. Still, the faintest star seen serves as a measure of how dark the sky is at different eclipses.

A planet ("Vulcan") lying nearer to the Sun than Mercury was expected during the 19th century, because of an unexplained acceleration in Mercury's orbit. At the 1878 July 29 American eclipse, Lewis Swift and James Craig Watson (discoverers, respectively, of many comets and asteroids) were convinced they saw the intra-Mercurial planet. But it was never confirmed, and the advance of Mercury's nodes was later exactly explained by Einstein's general theory of relativity.

Comets, swinging through the inner ends of their orbits, are genuinely likely to be found. Several of the bright comets of the nineteenth century, discovered by many people simultaneously, bear the name Eclipse Comet.

After-phases

Time's up! The colors of sunset on the *east* have swung around to become the fresh colors of sunrise on the *west*, where the umbra thins. The events recur in symmetrical order, beginning with another chance to glimpse the chromosphere on the Moon's opposite side, then the explosive reappearance of the Diamond Ring. The name was first used by spectators watching the re-emerging Sun, at the New York City eclipse of 1925 Jan. 24. In Canada on 1979 Feb. 26 Fred Schaaf saw, in a sky presumably misted with ice needles, the reappearing Diamond surrounded by two concentric cloud coronas—sharp, purple-red, and so small that the inner one crossed over the Sun's corona.

Another chance to look for the shadow bands, another chance to see the whole umbra rushing away. But human beings, unlike nature, are not patient with symmetrical stories, and do not generally wait through much of the closing partial eclipse.

A lunar eclipse: 1989 August 16/17

The Moon came to its New Moon position—directly opposite from the Sun—less than two hours before reaching its ascending node across the ecliptic. This meant that it passed nearly centrally through the Earth's shadow, spending an hour and a half totally inside the umbra.

This happened around the hour of 3 in Universal Time—3 hours after midnight on the zero line of longitude that passes through Greenwich. In eastern North America it was 5 hours earlier, that is, around 10 p.m. by natural time, which in summer becomes translated into the unnatural "Daylight-Saving" time of 11 p.m. The Moon, therefore, was sloping up the sky. It began to encounter the shadow when they were low in the southeast; the climax came higher toward the south; the watchers had mostly gone home or gone to sleep as the Moon freed itself from the shadow and reached its highest, in the south.

This was a lovely eclipse in golden August weather. I was lying on a southeast-facing grassy hillside that dropped away into woods. A single wisp of cloud perversely hid the opening stages; then the Moon emerged, became a black orange-edged droplet, and we watched for four hours.

An eclipse of the Moon is an event more leisurely and serene than one of the Sun; a smooth-topped hill rather than a giddy peak.

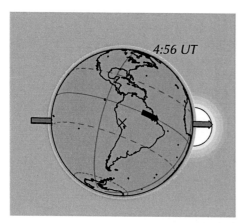 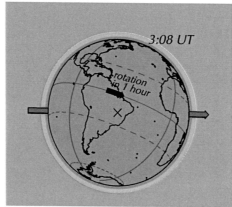 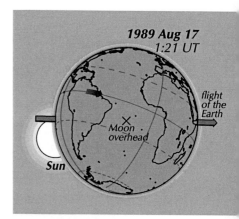

Views from the center of the Moon toward the Earth and Sun at the beginning, middle, and end of partial eclipse on the Moon. These show the parts of the Earth from which the eclipse could be seen. In the first picture, the U.S.A. is coming into view on the left; the Sun there has just set and the Moon is rising low in the southeast. Meanwhile Europe, on the right, is moving out of view: there, only the beginning of the eclipse is seen, because the Moon is about to set and the Sun about to rise. In the last picture, the U.S. has reached the midline of the globe: it is midnight, and the still partly eclipsed Moon is at its highest, though, because it is southerly (opposite to the summer Sun), it is still seen at a fairly low angle.

The thin ring of light around the Earth represents sunlight refracted through the atmosphere, which gives the reddish color of the eclipsed Moon. Its variable tone is due to clouds and pollution in the sunrise and sunset regions of the Earth at the time.

Movement of the Moon through the Earth's dark umbra and almost imperceptible penumbra, as seen from the center of the Earth. The umbra and penumbra are of course not visible except where they fall on the Moon. Arrows show the motion of Moon and shadow over a span of 8 hours. At this time in late summer, the Sun is in Cancer, and the full Moon opposite to it is traveling in the southern constellation Capricornus, about 2½° north of the star Delta Capricorni (Deneb Algiedi, "tail of the goat"). As seen from a north-hemisphere locality, Moon and shadow are displaced up to a degree southward, so that the crossing of the ascending node on the ecliptic happens later, and Mu Capricorni may be occulted.

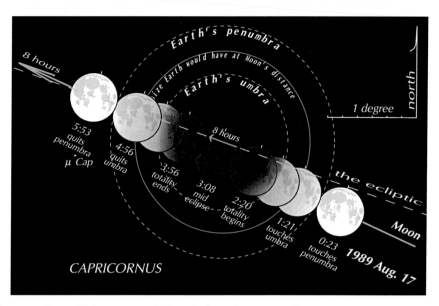

One of the most beautiful moments is just before the onset of totality: the Moon's last bright blue-gray sliver is about to be consumed by the crisp edge of the richly brown umbra.

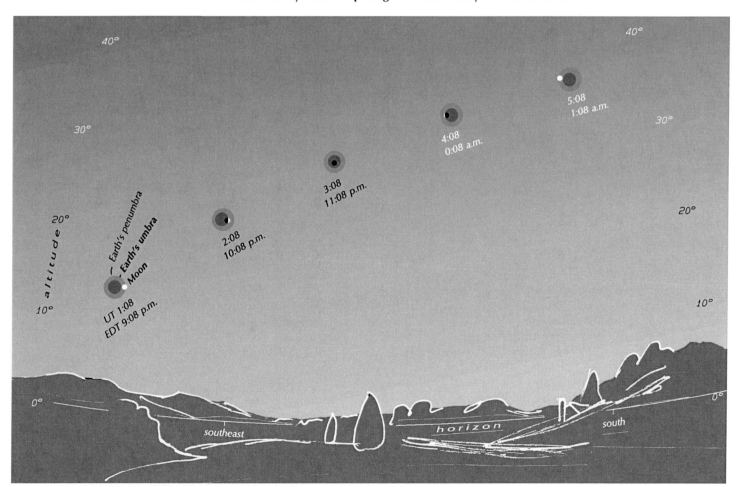

This is the actual sky scene for a particular locality, in this case the eastern U.S.A. (latitude 40° north, longitude 75° west). Moon and shadow are shown at intervals of whole hours from the central time of the eclipse. UT is Universal (Greenwich) Time; EDT is Eastern "Daylight-Saving" Time. The Moon moves eastward (leftward) across the shadow and both are progressing eastward from day to day; but the much more obvious diurnal motion—that is, the rolling of the Earth—carries Moon and shadow rapidly in the opposite direction, from their rising in the east to their culmination in the south about 5 hours later.

A partial solar eclipse: 1989 March 7 (Alaska)

The Moon passed through its ascending node at about 23 Universal Time on March 6 and had been climbing for nearly twenty hours before it crossed the Earth-Sun line. Thus it was already well to the north, and only the belly of its shadow brushed over the Earth. The part of the Earth thus brushed was mostly well forward of the north pole, down in the Pacific direction, for two reasons: because the shadow was climbing northward (the less important reason, the angle of its climb being only 5°); and because this was in March, so that the Earth was traveling in the attitude characteristic of that season: with its north pole tilted backward, at about 20°.

The north pole of the Earth in an *ecliptic* sense has to be measured from the center of the Earth perpendicularly up from the ecliptic plane; it is, in fact, a moving point in the Arctic Circle, where that intersects the terminator or day-night line. At the moment of our picture the terminator is slicing through the Bering Strait, and the ecliptic pole is just north of there, between Alaska's eyebrow and nose. Because of the shadow's 5° northward climb, the point of maximum eclipse was 300 miles south of this (just off Alaska's lower lip). For it was here that the umbra, the needle-like core of the Moon's shadow, swept closest. At this time the Moon was only 56 Earth-radii away, almost its nearest. Therefore the umbra easily reached past the Earth, and would have yielded a total eclipse if only it had aimed lower. But it passed 670 kilometers (420 miles) overhead. Day became darkest in the Bering Strait, or Alaska's side of it (the Russian side was darker, over the edge in night). The Sun became a thin bowl or boat, five-sixths eaten away from above. But at the point where the eclipse was darkest of all—where the heart of the shadow came closest—the mostly-eclipsed Sun was exactly on the horizon. It was rising, a boat with the horns of its prow and stern piercing into view first; but at an extremely low angle, floating along the southeastern horizon. This eclipse was a strange sight, as are all the eclipses and the other glancing phenomena of the far north.

Partial eclipses, if by that we mean all-partial eclipses—eclipses which do not become more than partial—are all phenomena of the Arctic and Antarctic. Eclipses that pass nearer to the Earth's center are partial in their opening and closing stages, but are named annular or total because of their middle phases. But the area over which even this all-partial eclipse was visible extended almost down to the equator. In fact the first touchdown of the outer shadow was east and slightly south of Hawaii. But in Hawaii and most of the U.S. and Canada and Greenland the eclipse was barely noticeable. The vast outer stretches of the penumbra are light; as seen from them less than half of the Sun is hidden, and the day hardly seems to darken.

This and the following views of the globe are from viewpoints out in space, at various angles but all at a distance of 10 Earth-radii from the Earth's center. An Earth-radius is 6378 kilometers or 3963 miles. (Geosynchronous satellites hover over the equator at a height of about 6½ Earth-radii, so you might see one pass below you.) Because of this finite distance, slightly less than half of the globe is seen. The Moon is about six times farther away behind you, back along the direction from which its shadow comes. The Sun, casting the shadow, is nearly 400 times farther in that direction than the Moon. The umbra, or total shadow, is shown as if it were a solid cone. Of the penumbra, which continues outward to infinity, the hazy flanks are shown, only as far as the Earth.

As further indications of scale (always measured in Earth-radii), the rotation-of-the-Earth and flight-of-the-umbra arrows are 1/100 thick; the latter is 1/10 wide (and about 1/2 long, but varying with the speed of the Moon).

The ecliptic, or plane of the Earth's orbit, is the plane in which lie the flight-of-the-Earth arrow and Sun-overhead arrow. These would meet at the center of the Earth, which is underneath the small cross in the middle of the picture. The Moon and its shadow do not travel in or parallel to this plane but at a 5° angle to it; at the time of this eclipse they are ascending northward from it. The interplay of these various movements—the thousand-miles-per-hour rotation of the Earth, the faster movement of the Moon past it, and the far faster orbital travel of Earth (and Moon) in the opposite direction—together with the spheroidal shape and tilted position of the Earth, results in the compound curves of the shadow's movements across Earth's geography.

Also passing through the center of the Earth is the "fundamental plane," used in all eclipse calculations. It is perpendicular to the axis of the Moon's shadow. It is nearly but not quite the same as the plane of the terminator or day-night boundary around the Earth (which is perpendicular to the sunlight).

The print of the Moon's penumbra, or partial shadow, on the Earth's surface is shown as a gray area, darker toward its center, where it is closest to being total because less of the Sun can be seen. Sometimes the outlines of other posi-

tions of the penumbra are drawn, at hourly intervals; sometimes, for the sake of less clutter, just the north and south limits of the whole area within which partial eclipse can at some time be seen—these lines are the envelopes of the positions of the penumbra.

You may be impressed by the vastness of the penumbra and the needle-like thinness of the umbra. How large is the Moon which throws them both? Its radius is about halfway between theirs: just over a quarter of the radius of the Earth.

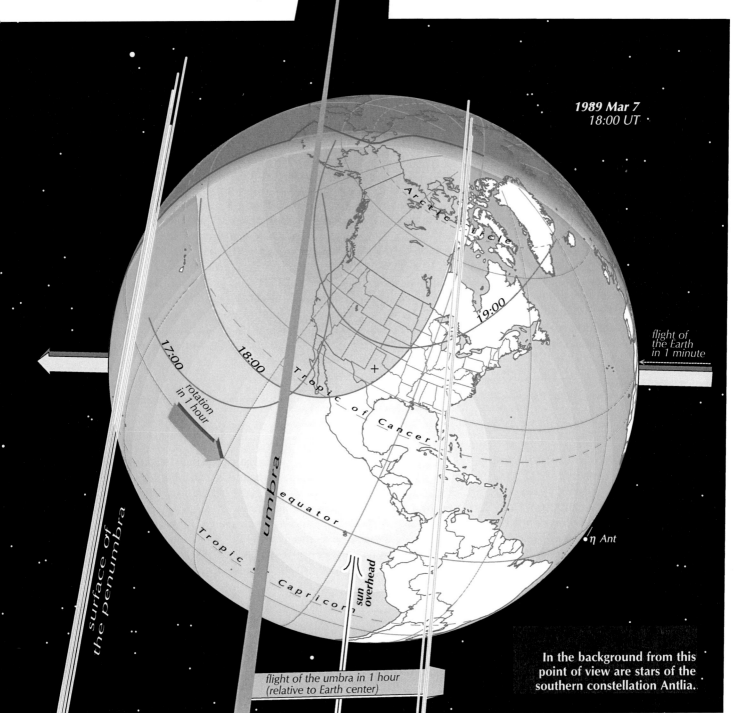

1989 Mar 7
18:00 UT

flight of
the Earth
in 1 minute

flight of the umbra in 1 hour
(relative to Earth center)

In the background from this
point of view are stars of the
southern constellation Antlia.

An annular eclipse: 1994 May 10 (cross-U.S.)

This was the only central eclipse (annular or total) to cross the contiguous United States between 1984 and 2017. It covered as much as possible—perhaps a seventh of the area—by making a wide path and laying it diagonally from west Texas (where I saw it) to Maine. But the main reason for the path's width was that the eclipse was fully annular, which was in turn because of the Moon's large distance at the time.

The wideness of the path was increased because the shadow passed north of center, where the Earth's surface sloped back. Notice that there was only the barest "northern limit of partial eclipse" (actually beyond the rotational north pole); only for a few minutes around the eclipse's mid-time was the penumbra wholly on the Earth.

The mean Moon distance is 60.3 Earth-radii (from the center of the Earth), which is greater than the length of its umbra plus an Earth-radius; therefore the umbra more often than not stops short in space. So there are rather more annular eclipses than total ones. Also, the patch within which an annular eclipse is seen can be larger-which implies that the path of annularity can be wider than a path of totality—and annular

Earth penumbra umbra **Moon**

15:00 UT

eclipse can be seen from one spot for a longer time: up to more than 12 minutes.

On this occasion in 1994 the Moon had passed its apogee only a day before, so it was at a nearly extreme distance of 63.6 Earth-radii. The umbra came to its end 4.5 radii above the Earth's center. The eclipse's midpoint was over the junction of the states of Indiana, Michigan and Ohio, 11 minutes past the moment chosen for our picture. By that time it was noon: the antumbra and the "Sun overhead" arrow, sweeping in opposite directions, arrived on the same north-south line of longitude. The Sun, seen 66° up in the southern sky, was covered by the silhouette of the Moon except for a shining ring barely more than one minute of arc wide, or 1/14 of the Sun's radius. This state of annularity lasted for about 6.2 minutes: the Moon detached itself from the right edge, swam slowly across the round pond of bright light, and touched on the left, breaking the bright ring. The width of the path, or of the patch of annular shadow moving along it, was about 230 kilometers (142 miles); in those same 6.2 minutes the forward edge of the patch left the observer and the rearward edge reached him.

Backward and forward along the path, the fit was slightly less tight—the visible ring

| | 10 | Earth-radii | 20 | 30 | 40 | 50 | 60 | 64 |

17:00 UT

north pole

1994 May 10

broader—because the Moon was more distant. The duration, however, as in a total eclipse, was longest at the center because that was where Earth's rotating surface was moving in the same direction as the shadow. To the north and south, mere partial eclipses were seen: the Moon from California (north of the path) covered the south of the Sun; from New York (south of the path) it covered the north.

This was not the annular eclipse of the very longest duration or widest path, for when the Moon is at apogee its distance can be as much as 63.8. The annular eclipse of longest duration in at least the 20th to 25th centuries was on 1955 Dec. 14, balancing that year's June 20 total eclipse which was also the longest in several centuries.

Plan of the eclipse from north of the ecliptic plane. Scale is true for Earth and Moon and the distance between them. The Sun would be about 59 metersto the right. But the distance between successive positions should be about 2.5 times greater. Each hour, Earth travels 8½ times its own width. The Moon is traveling with it, but, at New-Moon time, slightly slower. Its distance on this occasions makes its umbra fall short of Earth.

19:00 UT

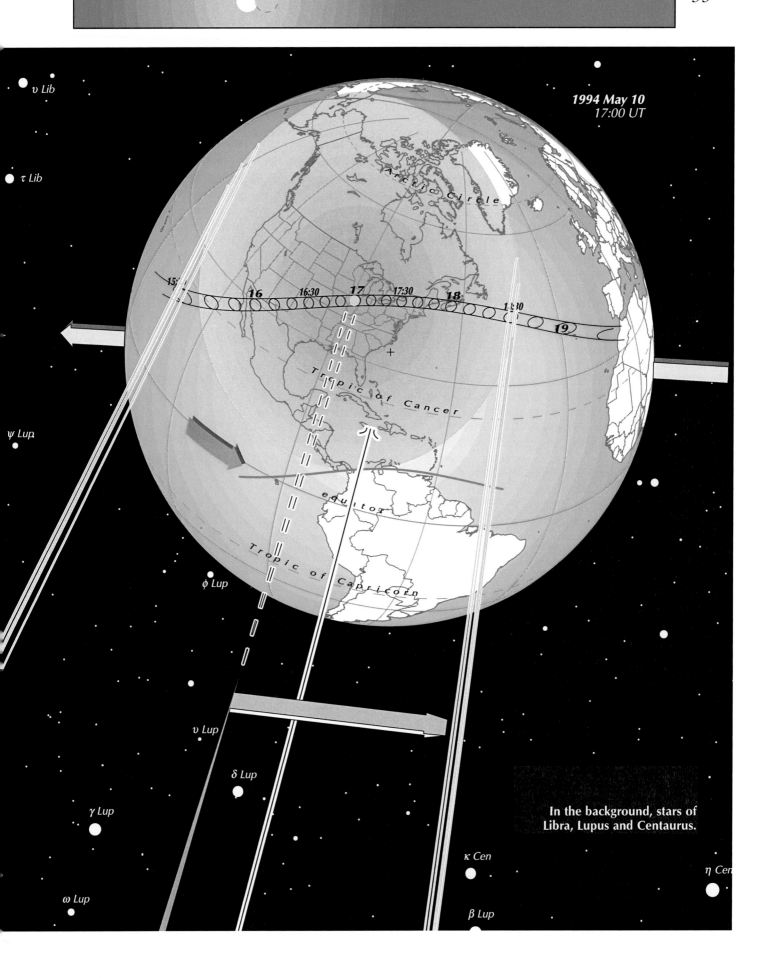

υ Lib

τ Lib

1994 May 10
17:00 UT

Arctic Circle

15:

16   16:30   17   17:30   18   1 30

19

Tropic of Cancer

ψ Lup

equator

φ Lup

Tropic of Capricorn

υ Lup

δ Lup

γ Lup

In the background, stars of
Libra, Lupus and Centaurus.

κ Cen

η Cen

ω Lup

β Lup

A broken-ring eclipse: 1984 May 30 (Greenville)

Imagine the Moon, each time it moves between us and the Sun, passing successively closer: its umbra would probe nearer; the imaginary antumbra continuing it would grow shorter, with a smaller cross-section on the ground; thus its track becomes narrower, the duration within which the annular eclipse is seen becomes briefer, and, most important, the Moon's apparent size grows nearer to that of the Sun—the fit becomes tighter. The ring of Sun left visible around this Moon takes on thread-like thinness. At any one place it is seen as a complete ring for only seconds before it is broken.

Thus an annular eclipse can approach, at its midpoint, the threshold to becoming total. This happened at the 1984 eclipse, the last central eclipse to cross the United States until 1994. Entering by the Mississippi delta, it followed the southeastern Piedmont, to leave by Chesapeake Bay, making a fine seam by comparison with the wide tramway of the eclipse almost parallel to it ten years later. It was to me a doubly strange eclipse because it was the only one for which I did not have to travel: I and many others watched it from the grassy expanse of a cemetery inside the South Carolina city of Greenville, where I was living.

The central phase of such an eclipse reaches the stage of Baily's Beads and there hovers, before retreating. The eclipse cannot be called total, since there is always some part of the Sun's photosphere visible; so it counts formally as annular. Yet in its middle phases it is not truly annular either, if that means that a whole ring remains visible: the ring breaks, it may have more breaks than bright links—it shatters into beads. No sooner has the last crescent been reduced to beads along the Moon's forward edge, than new beads begin to be born at the other edge. The smooth circle of the Sun and the jagged outline of the Moon, with almost exactly the same radius, are interlaced. The Moon is like a coin with a milled edge pushed into a hole of light. The beads flicker in and out of existence, with greatest rapidity and complexity at the north and south limits. From timing these "beading events" much can be learnt about the precise figure of the Moon and diameter of the Sun.

We might describe such an eclipse by saying that the Moon is almost exactly at the length of its own umbra from the nearest part of the Earth. The tip of the umbra's cone just touches, or almost touches, Greenville. Actually this forces us to see that the umbra cannot be quite a simple cone, though it is not easy to see what it is. Defined as the volume of space from within which none of the Sun's photosphere can be directly seen, it must be a sort of fluted cone, a cone grooved by channels along its sides, streaming from the valleys around the Moon's rim. Near the Moon these channels are proportionately very small, but toward the tip, where they are as large as they started but the umbra has become very narrow, they must cut into it from all sides. Finally they must

shear its end away, as when you sharpen a pencil point with a knife. This eroded umbra-tip must end farther back in space than the theoretical umbra cast by a circular Moon. The last needle of the umbra is not merely whittled away but replaced by the intersecting grooves, from within each of which one can see a bead of light coming straight from the Sun.

Or we could ask what outline is traced on the Earth's surface by the rays that have grazed the Moon's edge. Suppose the Moon is circular. At a total eclipse, where the umbra cone is cut off by the Earth, the ground-print is a small circle, defining the patch of totality. If the cone is pulled back (the Moon more distant), the circle shrinks. At an exactly borderline eclipse, the rays meet at a point on the ground. At an annular eclipse, the rays cross in space; the circle turns inside out and becomes a negative circle around the patch where the annular eclipse is seen. Now, the mountains on the Moon make the circle bumpy. Relative to a large circle these bumps are small. As the circle shrinks, the bumps stay as large, so that they dominate and the "circle" is a highly irregular amoeba. As *this* passes through itself and turns itself inside out, it becomes an almost unimaginable tangle. This expresses the fact that, standing on the ground inside this map, you see sunlight blocked by some mountains and coming through some valleys all the way around the Moon's edge; you are in a shadow overlaid by several bays of light.

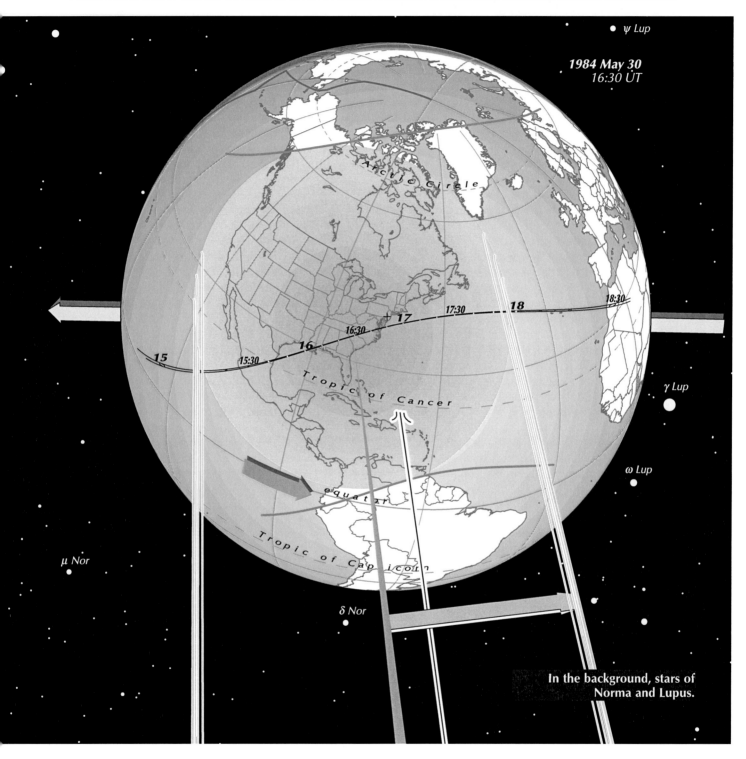

In the background, stars of
Norma and Lupus.

On this date the umbra of the Moon's average surface (depending how you define that) stopped maybe 51 kilometers short of the Earth; but the shadows of some mountains would have met under the surface, whereas the anti-shadows of the valleys met perhaps 1000 kilometers farther out.

This was not the annular eclipse most close to touching down and becoming total. The similar eclipse eighteen years (a saros period) before, on 1966 May 20, was even closer; the one preceding that, 1948 May 9, closer still. Then, the theoretical duration of the annular phase on the centerline was reduced to 1/10 of a second! The precursor eclipse to that, 1930 April 28, actually became briefly total in its middle.

An annular-total eclipse: 1987 March 29 (Gabon)

If the Moon passes slightly nearer still, its umbra reaches past the surface of the Earth, but not as far as the Earth's central plane. It ends—or would end, if it were made of steel—somewhere inside the Earth's day-side hemisphere.

On this occasion in 1987, the Moon was ascending from the south, and was due to reach the node on the ecliptic less than 6 hours after passing between Earth and Sun. So the eclipse was fairly central: the axis of the shadow passed south of the Earth's center by 3/10 of an Earth-radius. The northward slope of the track—only at the usual 5° to the ecliptic—was exaggerated in equatorial terms because of the Earth's attitude near the March equinox, with its north pole leaning backward. In the middle of any track that passes roughly across the Earth's middle, the surface is turning in about the same direction as the Moon's shadow flies, nearly half canceling the shadow's great speed. At the beginning and end of the track, the shadow's relative speed is much greater because the Earth's surface is angled away; hence the track curves to fit the shadow's eastward rush. The end of the track off the horn of Africa is visible in our picture; actually when the axis of the shadow reaches it, two and a half hours later, rotation will have carried it around out of sight to where the sunset terminator is now.

The Moon's distance at this date was such that its umbra ended about 0.15 of the way from the nearest dome of the surface to the Earth's mid-plane, that is, about a thousand kilometers underground. When the umbra, sweeping through space, first came into aim at a part of the Earth—Patagonia (southern Argentina)—it was five thousand kilometers too short, and the Patagonians saw the Sun rise (fairly exactly in the east) in annular eclipse. At the moment of our picture, the umbra has just struck the surface: the eclipse has become total. The track of totality runs almost entirely over ocean, passing only 25 kilometers southeast of the island of St. Helena. Close to here the eclipse is at maximum: the axis is at its nearest to the center of the Earth; the noonday Sun is at its closest to being overhead, just 21° northward up the 6° line of longitude; the umbra reaches its thickest—not very thick, only 5 kilometers—on the watery surface, and the total eclipse lasts its longest, only 12 seconds. Very shortly after reaching land again, almost exactly on the equator in the country of Gabon, the eclipse turns back to annular: the umbra's tip loses contact with the Earth.

An ec;o[se that is annular-total. sometimes called "hybrid," makes the narrowest overall track, since it hovers about point width. It crosses through point width twice, the tracks of annularity and totality both widening away from these crossover points: exaggerated, they should look like this:

Since the track is so narrow, on our scale it obscures the
very small elliptical patches of annular and of
total eclipse; plotted alone, they look
like this:

At each of the crossover points, the Moon fits over the Sun as exactly as it can.

You may wonder why, since the picture is at 12 Universal Time, the "Sun overhead" trident does not strike exactly on the 0° line of longitude. The reason is the "equation of time," or difference between the mean position of the Sun and its real position, which at this time of year is about 5 minutes.

Within the small class of annular-total eclipses, there is an even smaller (called by Jean Meeus "semi-hybrid"). They are annular only at one end of the track instead of both. Such was 2013 Nov. 3. The central track started as annular, in the Atlantic between

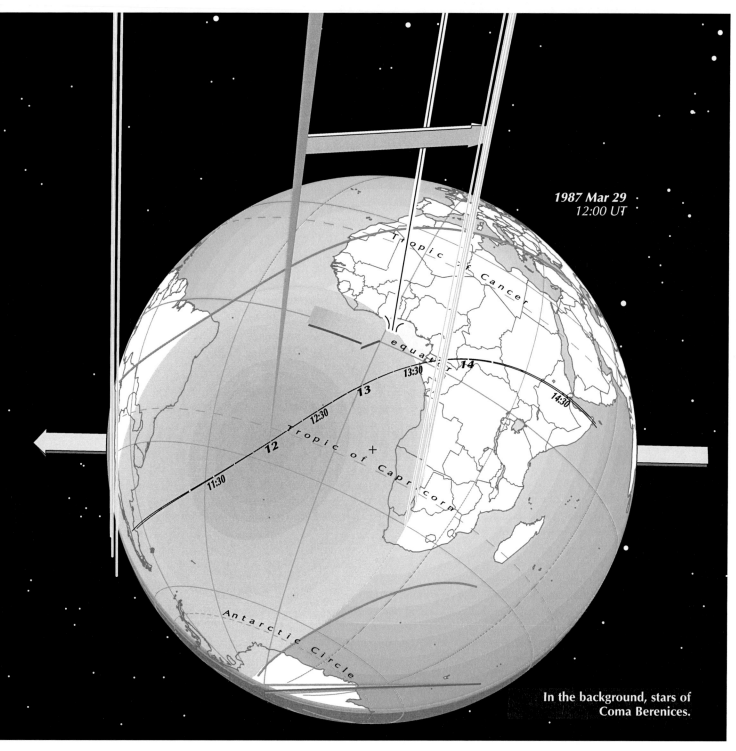

1987 Mar 29
12:00 UT

In the background, stars of
Coma Berenices.

Florida and Bermuda. It changed to total after only 15 seconds! At this moment the tip of the umbra hit the surface: the Moon (if we ignore its detailed shape) fitted over the Sun exactly; the duration of annularity ended at zero and that of totality began at zero. Then, the umbral cone digging (as it were) into Earth, the eclipse remained total all across the Atlantic and Africa to the end of the path on the coast of Somalia. The cone did not quite come out: it had probed slightly deeper, because during the eclipse the Moon had moved slightly near-er (its perigee was three days later).

The previous such case was 1854 Nov. 20, the next will be 2172 Oct. 17—a periodicity of 1 in 159 years.

A globe-skimming eclipse: 1986 October 3 (Iceland)

This is the converse (in some ways) of the partial eclipse of 1989 March 7. The Moon, coming to the end of its cycle, was *descending* southward toward the node; and it had not yet reached the node. So its shadow brushed over the north, but in a mirror-image way. At this northern-autumn season the Earth's attitude was north-pole forward. So the shadow brushed closest just past the topmost point, on the rearward side.

Because the interval between node and New-Moon times was slightly shorter—18 hours instead of 20—the shadow brushed slightly lower. Notice that the region of partial eclipse (of similar but reversed shape) reached this time to slightly below the equator. And now the axis itself, instead of missing by a few hundred miles, just intersected the globe.

However, whereas at the 1989 eclipse the Moon was much nearer than average, at 56 Earth-radii, this time it was only a little nearer than average: 58.65. This distance was only slightly more than the length of the Moon's umbra. The result was that the umbra reached almost exactly to the central plane of the Earth.

So this eclipse was borderline in two ways. The flight of the axis was almost tangent to the Earth: if the axis had been a wire, it would have removed a scab only a thousand miles long and up to a dozen miles deep, containing Iceland and a piece of the ocean south and west of it. And the distance along the axis to which the umbra reached was also borderline: statistically more likely was that it would reach far past, yielding a total eclipse, or stop far short, yielding an annular one, but it reached just to the scene of the activity: the umbra's very tip was involved.

The central eclipse started at 18:55.5 Universal Time, as an annular one. The point at which the axis struck the Earth was almost exactly on the Arctic Circle, which passes barely north of Iceland. In other words, this point, being also on the sunset-line, was the Earth's topmost point at the time: its ecliptic north pole. (From this fact, given the 5° descent of the Moon's orbit, one could calculate how many miles under Iceland the axis passed.)

Only 1.4 minutes later the umbra tip struck the Earth: total eclipse began, at a theoretical point. It proceeded for 17.9 minutes, the very narrow track buling Moonward in a strong curve and slightly re-widening in its middle. Then the umbra tip came out of the ocean again, the eclipse was again annular, and it ended 0.9 of a minute later— on the same sunset line, though this had twisted westward for twenty minutes. If you were able to water-ski down the sunset line at 1800 miles an hour, you would see the Sun sitting in the western horizon all the time, while the Moon's silhouette at first blocked its middle, then sank lower, then again blocked the middle, while shifting leftward:

To see the Moon higher, completely blocking the Sun, you would have to ski faster, along the curving track of totality:

The October waters between Iceland and Greenland do not have the best weather for seeing an eclipse, especially a low one. The best view was seen from an airplane. It did not fly over the surface track but many miles to the west, the track in the high atmosphere being displaced in the direction of the Moon.

At the small scale of the picture, it is scarcely possible to see the boundaries of the annular and total eclipse track, which (exaggerated) cross over like this, the crossings being the points where the umbra tip touches and the eclipses changes from annular to total and back

The shapes of the antumbral and umbral footprints are drawn at 1-minute instead of 10-minute intervals. The umbral ones should really appear like this: a set of narrow ellipses, becoming dots at the crossover from and to the annular phases.

This doubly borderline eclipse is sensitive to slight inaccuracies in calculation.

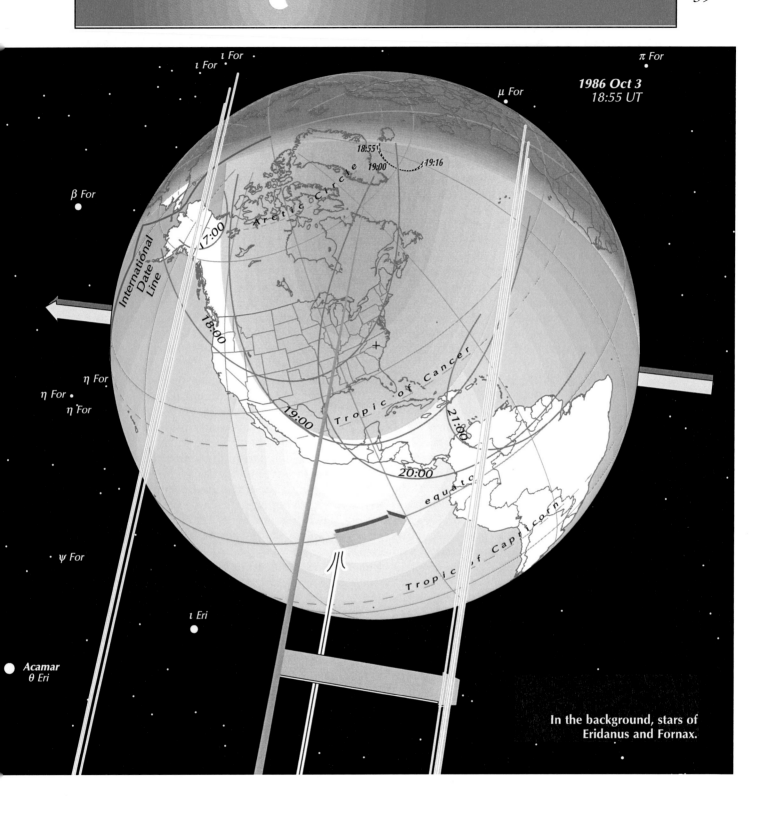

1986 Oct 3
18:55 UT

In the background, stars of
Eridanus and Fornax.

There was an unusual observation during this eclipse. Russell Eberst was tracking
satellites from Edinburgh in Scotland. The 10 that he saw between 19:34 and 20:36 UT
(beginning with the Russian Cosmos 1682, 1735, and 1674) were on average 1½ mag-
nitudes fainter than expected. He was over the night border of the Earth; they were at
heights where they should still have been in sunshine; but he realized that they were
inside the Moon's passing penumbra.

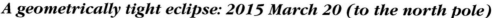

A geometrically tight eclipse: 2015 March 20 (to the north pole)

Timing determines geometry. This eclipse (1) almost coincided with the date of the March equinox, and (2) came at a nicely chosen moment in the Moon's slide toward the ecliptic plane.

(I could have said "narrowly chosen," but I'm not averse to *nice*: it still has its older sense of "precise" or "careful" alongside its over-used but useful sense of "pleasing.")

The Moon arrived between Sun and Earth (the New moment) 13 hours before the equinox, and 16½ hours before descending node. The equinox is when Earth is traveling with its north pole leaning maximally backward.

Picture Earth as an egg, and the Moon's umbra as a knife with which (left-handedly) you slice off the top of your egg. The knife strikes at point B; is deep in where the egg bulges out at F and S; and comes out at N. The decapitated cap is a circular dome, with center near E.

If the time before descending node had been longer, the umbra would have passed higher and missed Earth; if about 18 hours, it would have passed one Earth-radius from the center, and grazed point E. But because that span was 16½ hours, the cut was about 5% of the way to the center. This caused point N to be at the north pole.

The rotational north pole, that is. The point at the top of the Earth, in ecliptic terms, was E, which we can call Earth's ecliptic north pole. It is a moving point, as the Earth rotates. Because the date was the equinox, E was as far as possible from N: by about 23°, the angle of Earth's tilt. That is, it was on A, the Arctic Cirle. And, since E was the center of a circle, B was about the same distance on south. Not exactly, because of the slanting motion of the Moon during the 70 minutes of the path.

E was in the middle of Greenland; B, south of its tip. The path made a semicircle around Greenland, thus lay almost entirely over water. It passed between Greenland and Iceland on one side, Scotland and Norway on the other, touching only the island groups of F, the Faeroes (where I saw this eclipse) and S, Svalbard (where many serious observers went).

Because the umbra struck the surface at a low angle, and because the Moon was near (perigee was the day before, one of the year's nearest perigees), the part of the umbra that struck was not the narrow tip but relatively broad. So the relatively short semicircular path was more than 400 kilometers wide. The footprint of the umbra at any one time was a long ellipse across this path.

The penumbra was at least partially on the Earth, giving partial eclipse, for more than 4 hours, but the umbra's path was on the Earth for little more than one hour. Halfway along it, at 9:45 UT, total eclipse was at maximum, lasting 2 minutes and 50 seconds on the centerline, with the eclipsed Sun 19° above the southern horizon.

Toward the end, the path of totality kept curving, closer to exact north. Because the surface was curving away, the Sun was lower (in local afternoon), and the speed of the umbra over the surface accelerated. The centerline ended only 2/3 of a degree short of the exact north pole. So the two limiting lines at the sides of the path, normally called the "northern" and "southern" boundaries of the path, ended almost exactly on opposite sides of the pole.

The strange result is detailed by Jean Meeus in his *Mathematical Astronomy Morsels IV* (2007), pages 40-43. He wondered why he had not, in his *Morsels III* three years earlier, included this in a list of "large" solar eclipses at the poles (of which he had found between 2 and 10 per century).

The equinox was just a few hours after the eclipse ended. At the equinox, the north pole's six months of 24-hour night end and the Sun circles along the horizon. Therefore, throughout the eclipse the Sun was just below the north pole's horizon. Or was it? Geometrically, yes, but this applies to the Sun's center rather than its top; also, refraction, strongest at the horizon, raises the Sun's image by about 34´ (with noticeably flattened shape because of the rapid decrease of the refraction upward; probably distorted, also, by differently thick layers of atmosphere).

Therefore, at the north pole the *apparent* Sun *was* entirely above the horizon throughout the eclipse. The Moon dented its right edge (below center) at 9:22 UT; totally covered it for 1 minute 38 seconds, centered at 10:17; and quit its upper left edge at 11:12.

The peculiarity of the eclipse resulted from the coincidence of those two factors: its near-coincidence with the equinox; and the exact amount of its timing before the descending node.

```
Mar 19 19:30 UT Moon at perigee
Mar 20  7:41 start of partial eclipse      10:21 end of total eclipse
        9:10 start of total eclipse        11:50 end of partial eclipse
        9:38 Moon New                      22:47 equinox
        9:45 greatest eclipse       Mar 21  2:17 Moon at descending node
```

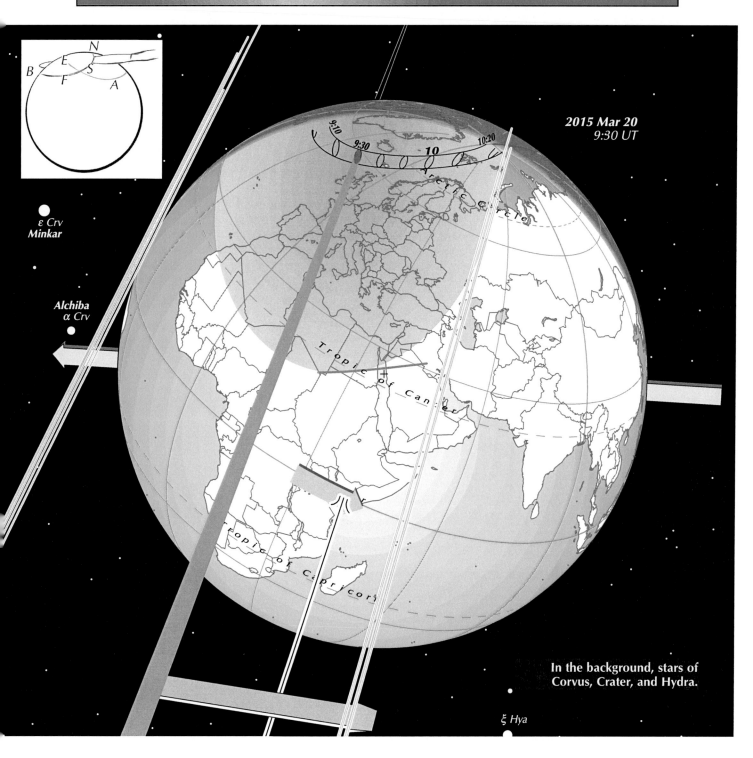

2015 Mar 20
9:30 UT

ε Crv
Minkar

Alchiba
α Crv

In the background, stars of
Corvus, Crater, and Hydra.

ξ Hya

Meeus points out further peculiarities about the edges of the path of totality, which are "northern" and "southern" in relation to the ecliptic but have become level with each other in relation to Earth's spin axis: "The remarkable fact is that, while the southern limit of the path of totality ends at local sunset . . . the northern limit ends at local sunrise . . . , at the other side of the pole."

One might say that the calculated lines for a solar eclipse's path of totality should spread slightly, both sideways and endways, at the two ends of the path by adding the (average) effect of refraction. When the Moon is high in the sky, refraction does not shift its silhouette on the ground, but when it is near the horizon it does, enough to make the difference sometimes between visible and not.

As far as I know, no one was at the north pole to watch as the Sun, theoretically below the icy horizon, bowled along above it and was blacked out by another horizon-bowling globe.

A high-latitude total eclipse: 1990 July 22 (Finland)

The Moon came aiming down to reach the descending node 14 hours after passing the Earth-Sun line. So the umbra cut across a little more than three quarters of the way up the globe. But with the north hemisphere tipped Sunward in its summer position, the track seemed more northerly, rising to a latitude only 13° from the north pole. The track formed a descending line in relation to the ecliptic (or, we could say, as seen from the Sun); geographically, it also trended generally southward, starting at 60° latitude and ending at 30°; however, again because the globe's head was bowed sunward, the first part ran steeply northward. Almost the only hopeful region for observation was the very beginning; many visitors went to Finland, welcomed by the active Finnish astronomy clubs.

At the gateway of the Gulf of Finland, the Sun rose in eclipse at 1:53 UT, which was 3:53 a.m. by Finnish clock time—an early hour, because northern summer nights are short. The umbra whipped across the 300 miles to the border of Soviet Karelia in little more than a minute, of course very flat to the horizon all the way. When an eclipse or any other celestial spectacle lies so low, it is very easily blocked by trees, mist, or a single far-off cloud. Some astronomers flew a special plane at 42,000 feet; even for them totality was only 80 seconds and some, operating their instruments, had no time to use their eyes on the eclipse.

The umbra got rapidly away through the White Sea into the Arctic Ocean and visited its remote peninsulas and islands, mostly blanketed with fog. Cloudy weather probably lasted until the eclipse crossed the Aleutian island chain at Atka. At the very end it came into open skies in the ocean north of Hawaii, and there was talk of seeing it from ships and planes there.

Kuun varjossa, "In the shadow of the Moon," says the caption to this photo from the Finnish magazine *Tähdet ja Avaruus.* And the horizon light from penumbral regions to north and south seems to define the cone of the umbra coming to us across the plains of Karelia and Archangel.

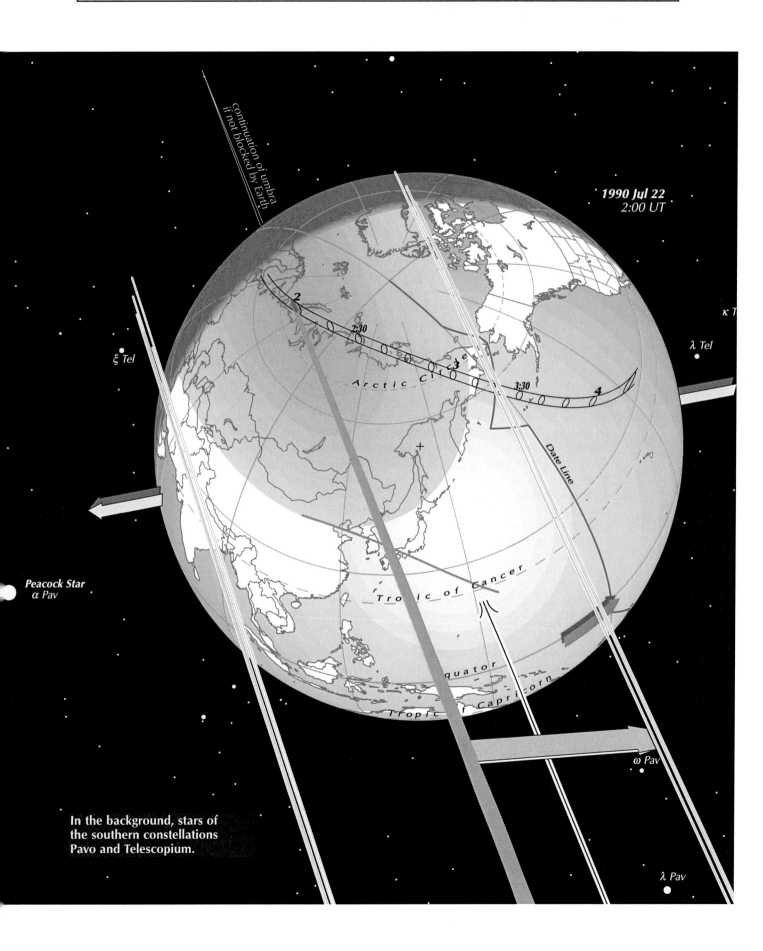

Continuation of umbra
if not blocked by Earth

1990 Jul 22
2:00 UT

κ T

λ Tel

ξ Tel

2

2:30

3

3:30

4

Arctic Circle

Date Line

Tropic of Cancer

Equator

Tropic of Capricorn

Peacock Star
α Pav

ω Pav

**In the background, stars of
the southern constellations
Pavo and Telescopium.**

λ Pav

An ideal eclipse: 1991 July 11 (Hawaii and Mexico)

A bull's-eye eclipse! Or a kebab: Sun, Moon, and Earth on a skewer.

Its distinction was that it hit the center of the Earth. Actually, the axis of the shadow passed a mere 16 miles south (the quantity called gamma being –0.004, or 0.4 percent of Earth's radius). It was the most central eclipse in nearly two thousand years: since the eclipse of 45 Aug. 1 (overhead between Sudan and Arabia), for which gamma was –0.0011 or about 4 miles.

Why then is the umbra not shown striking the equator? Because in northern summer, around the June 21 solstice, the Earth has its north hemisphere tipped toward the Sun. Only at the March and September equinoxes is a point on the equator the Earth's midpoint as seen from the Sun.

A central eclipse is geometrically simplest. The Moon came sliding down at a time such that it would arrive simultaneously (to within 3 minutes) at its descending node and at the New Moon position. The front of the penumbra struck the front or sunrise surface of the Earth; partial eclipse began. An hour later, the umbra struck; the path of total eclipse began. An hour and forty minutes later, the umbra crossed the summit of the Earth's dome; it pointed straight down to the Earth's center; it stood vertically, so that people within it lay on their backs to look at Moon and Sun stacked over each other in the zenith. The umbra was at its shortest and thickest, and moved most slowly in relation to the turning surface, so that totality lasted 6 minutes and 58 seconds.

An hour and forty minutes later, the umbra whisked off the edge of the Earth that was receding into night; total eclipse ended. Another hour later, the rear surface of the penumbra left the rear surface of the Earth; partial eclipse ended.

The timing in Earth's rotation determined the geography. The western or morning half was ocean except for two pads of land, each smaller than the umbra's footprint: the "Big Island" of Hawaii (with great observatories on its peaks) and the tip of the peninsula of Baja ("Lower") California. On both, the July climate was expected to be mostly cloudless, so these were the two places to which the most thousands of eclipse-chasers headed. Then, overheadness on the mainland coast. The afternoon half of the track ran almost all over land, much of it densely populated—but Mexico, Central America, and Colombia were under their rainy season. At its extreme end the track emerged south of the Amazon into a dry south-equatorial winter; but here the Sun was very low, about to set.

As it happened, clouds covered Hawaii except for its privileged peaks; and the best luck was at the center of the track, on the Mexican coast. Morning clouds built up as usual on the mainland—then melted back as the approaching shadow cooled them.

More detail about this eclipse was in my *Astronomical Calendar 1991*, and there will be many references to it later in this book, which was first composed in anticipation of it. I wrote myself a 16-page description of my journey (by bicycle) to see it, at a Mexican fishing village, Sayulita.

On the next two pages are pictures of the sky during total eclipse. The charts are centered on the zenith, so that the horizon becomes a circle. The scale is 1.5 millimeters per degree of altitude. The dashes along the ecliptic are 2° long, and the ticks on the celestial equator are an hour of right ascension apart. The Moon's silhouette is twice its real ½° width.

From Hawaii, the Sun was 20° above the horizon (and the huge mountain Mauna Loa, 14,000 feet high, stands to about 7° as seen from the western coast). From Mexico, the Moon and Sun were almost in the exact zenith, 90° above all points on the horizon. Lines drawn from opposite points on the horizon—northeast and southwest, northwest and southeast—would intersect at the zenith. The position of the eclipsed Sun in relation to the zenith is sensitive to where exactly we are in Mexico: the picture is plotted for Tuxpan, 6 miles southeast of the subsolar spot, so the Sun is just slightly north of the zenith.

Some stars and planets became visible in the darkened daytime sky. The most likely are the brightest; here are those shown, listed in order of their "magnitude," the astronomical measure of brightness:

VENUS	-4.5	MERCURY	-0.1	Betelgeuse	0.8	Adhara	1.5
JUPITER	-1.8	Capella	0.1	Aldebaran	0.9	Castor	1.6
Sirius	-1.5	Rigel	0.1	Pollux	1.2	MARS	1.8
Canopus	-0.7	Procyon	0.4	Regulus	1.4		

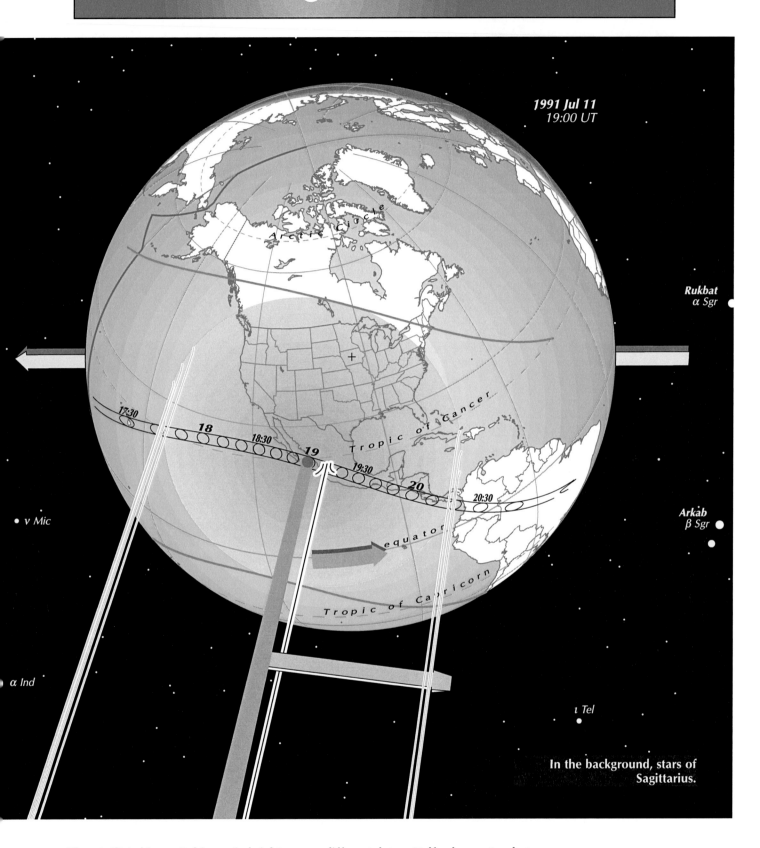

1991 Jul 11
19:00 UT

Rukbat
α *Sgr*

ν *Mic*

α *Ind*

Arkab
β *Sgr*

ι *Tel*

In the background, stars of
Sagittarius.

17:30 18 18:30 19 19:30 20 20:30

Planets (listed in capitals) vary in brightness on different dates. Half a dozen stars between
the brightnesses of Castor and Mars are not shown. Canopus is off the page to the right
of Sirius; a deep southern star, second brightest in the whole sky, it is just rising in the
south-southeast for Hawaii, and is 14° high in the south for Mexico. Whether you actually see any of these at a given eclipse depends also on how far they are from the glow of

the Sun's corona and from the horizon, how quickly your eyes adapt, and the sky conditions. It happened that four of the five bright planets (omitting only Saturn) lay within 41° east of the Sun on this date, so that they had been visible low in the evening sky; in fact they had been putting on a rare display, a "grand conjunction," for several months, climaxing around June 18 when Venus, Mars, and Jupiter were packed within a 1.8° circle. In the mornings, rising behind the Sun, they had not been visible. But now suddenly the eclipse allowed them to appear beside the Sun by day. At eclipse time in Hawaii they were still below the horizon, but in Mexico an hour and a half later they had leaped halfway to the zenith. Venus, brighter than all other planets and stars, was the likeliest to be visible, in fact it is sometimes spotted in the deepening blue of the sky as much as half an hour before totality. Regulus appeared between Venus and Mars. This gave a sense of the third dimension: in the foreground, the Moon; behind it in the solar system were successively Venus, the Sun, Mercury, Mars, and Jupiter—they were about 200, 400, 450, 1000, and 2600 times farther than the Moon. And behind those lay Regulus, one of the stars near enough to be visible to the naked eye—at 85 light-years or about 2,250,000,000 Moon-distances.

The Milky Way (composed of stars mostly hundreds of times farther than Regulus) is not visible in an eclipse sky, but is shown for orientation. To those who know the constellations, it gives a sense of which part of the universe we are facing.

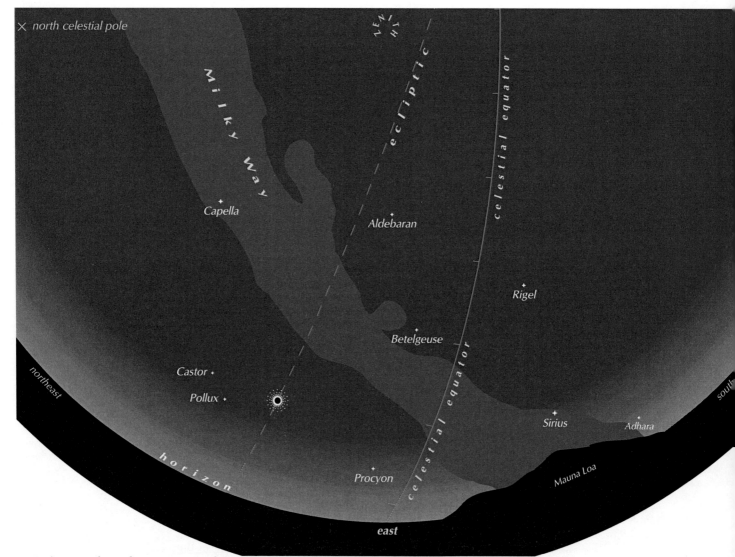

Facing east from the west coast of Hawaii at 17:28 Universal Time (7:28 Hawaii standard time, 8:28 Hawaii "Summer" or "Daylight-Saving" time). The eclipse is only 20° above the horizon.

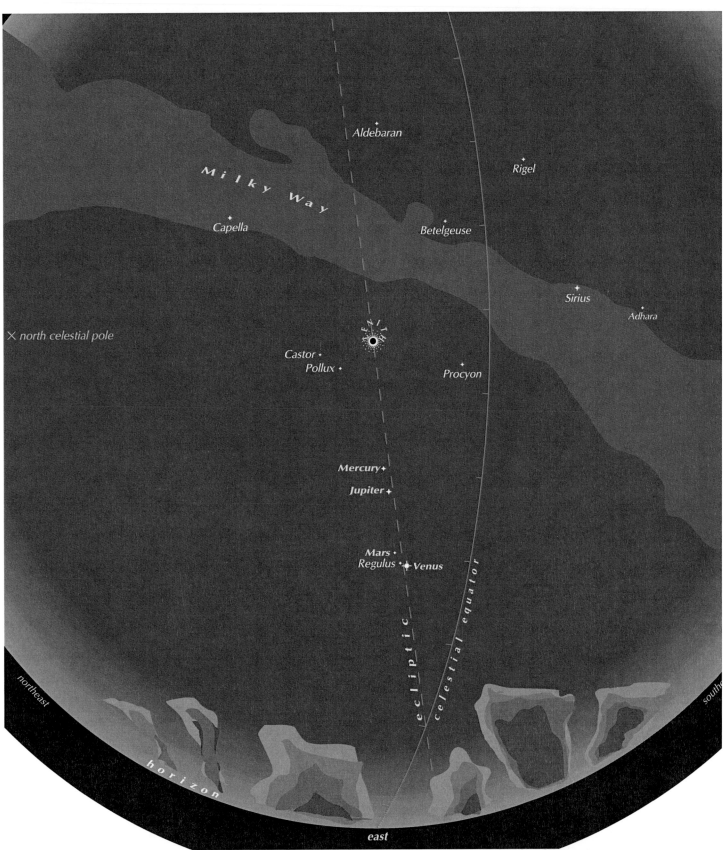

Facing east from the west coast of mainland Mexico at 19:00 Universal Time (12 noon in local and in western Mexican standard time). The eclipse is now at the zenith, and below it four planets have climbed into view. Thunderclouds that threatened to arch overhead have been suppressed by the cooling of the air.

Further aspects of the 1991 eclipse

Diagram (at a scale of 10 centimeters to 1 degree) showing how the Moon passed across the face of the Sun on 1991 July 11, as seen from the centerline at Tuxpán in Mexico (latitude 21° 55´ north, longitude 105° 18´ west). The Moon moves from west to east across the Sun. Actually we choose to show the Moon standing still and the Sun's position relative to it (at 5-minute intervals), moving from east to west. This is more like what you perceive during the central moments: the Moon materializing as a black circle, and the Sun disappearing behind its east limb, then reappearing at the west. This way, we can show the progressively thinner and shorter crescents of remaining Sun. Notice the last thin crescent on the east, about to dwindle further and break into Baily's Beads and then the Diamond Ring; and the first thin crescent appearing on the other side, having just dazzled you with the second Diamond Ring.

Seen from the centerline, the Sun dwindles symmetrically and spends the longest time behind the Moon, granting the longest chance to marvel at the gauzy details of the corona.

Tuxpan
horizon 90° below

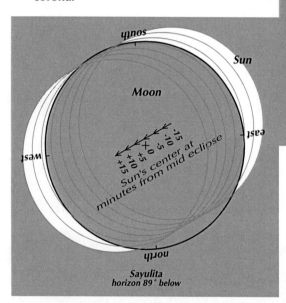

Sayulita
horizon 89° below

But this is how things are different as viewed from near the edge of the track of totality, instead of from the centerline. The members of Tom Van Flandern's Eclipse Edge Expedition chose to be at the fishing village of Sayulita (20° 50´ N., 105° 30´ west)—only 4 or 5 kilometers inside the southern limit. The chord of totality was shorter, but not as much so as you might think: 1 minute 44 seconds, as against 6m 58s on the centerline. The great difference was that the crescent of the Sun dwindled asymmetrically and over a longer time, especially along the Moon's southern curve. Notice how the Sun crescents dwindle more gradually and sidle around the southern curve of the Moon. They broke into a longer and more gradual succession of Baily's Beads between the mountains along this edge—which is rougher with mountains than the northern. Other edge phenomena such as the shadow bands were enhanced and prolonged; and the red chromosphere, which is narrower than the last crescent shown, instead of flashing into and out of view at subliminal speed, was in view *throughout totality*, wrapped around the Moon's south pole.

The pictures are oriented so that the nearest point on the horizon is below. However, at Tuxpán the eclipse took place almost exactly at the zenith, and at Sayulita barely more than a degree north of it, so one's feet could be in any direction. I had intended to retain clarity by facing south; but as the Sun ascends in the east throug the late morning you watch it that way, and tend to continue doing so as the excitement mounts and arrives overhead; so in this sense it would be more natural to show east at the bottom. To the east was the mountain rim of Mexico and a nearby line of palm trees; as the morning became hotter a line of thunderclouds built up from the mountains, began to tower over the palms and threaten the eclipse, then shriveled away as the mustering shadow cooled the air.

The photosphere is the normally visible surface of the Sun, 696,000,000 kilometers in radius. The chromosphere is the inner atmosphere, only about 2,000 km thick. The corona is the outer atmosphere, which fills space outward past the Earth.

This is a schematic graph of brightness across the Sun. The corona even at its inner edge is about 500,000 times fainter than the photosphere, and it plunges away downward.

The lower lines show how the Moon, sliding across the Sun from right to left, cuts out the tower of brightness and excavates a black trench in its place. The dotted line shows the situation just before totality. A mere needle of photospheric brightness is left. At the dashed line, the whole tower has fallen, leaving only its buttresses—the slopes of the corona. For an instant, only, the chromosphere alone is exposed before it too is covered.

The bottom of the trench does not go down for ever. Faintly on the Moon can sometimes be seen and photographed the earthshine: sunlight reflected back from our planet, and showing 9,000,000,000 times fainter than the photosphere.

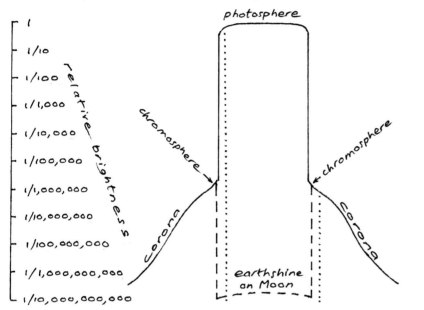

To open your lecture on how you experienced the great eclipse, stand a globe like this: place of noon eclipse (Mexico) facing audience; globe tilted toward the line of longitude west of this by as many degrees as the days since June 21 (thus for July 11: 135° W.).

"This is our Earth, 8000 miles wide, reduced to a globe 12 inches wide. It's set for July—north-hemisphere summer, so the north hemisphere is tilted toward the Sun. This is the Moon, a quarter the width of the Earth, reduced to an orange, 3 inches wide. It should be 10 yards away—one, two, three ... The Sun should be nearly 2¼ miles away in the same direction, and should be 112 feet high! The Sun sends the Moon's shadow toward the Earth. And to show how the shadow moves across—that is, the pointed middle part of the shadow, which we call the umbra—ah, I spy just the thing!" Step across and borrow it from stooge. "The umbra, reduced to an umbrella."

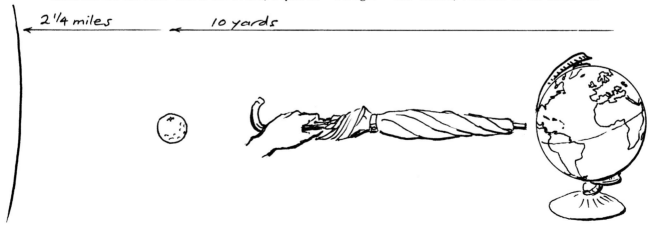

Eclipses of 2017

2017 is a year of the common four-eclipse kind: two eclipse seasons, each consisting (this time) of a slight lunar eclipse followed two weeks later by a greater solar eclipse.

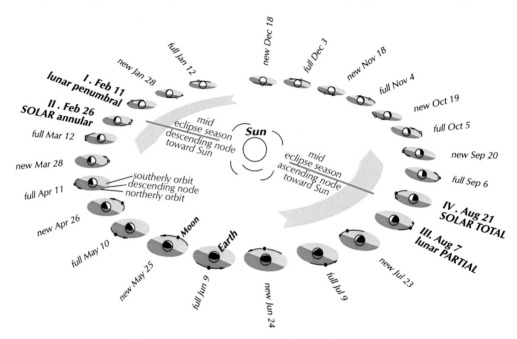

SCHEMATIC VIEW summarizing the year's eclipses. At each date of New or Full Moon, the Earth is shown with the Moon inward of it at New Moon, outward at Full. The plane of the Moon's orbit at the time is shown in blue, paler for the half lying south of the ecliptic. This plane gradually rotates backward. There is an eclipse if the Moon is Full or New when it is in the ecliptic plane, that is, close to the time it crosses the ascending node of its orbit or the opposite descending node. The black arrow is the Moon's course over 7 days. The view is from ecliptic longitude 270°, latitude 30°. Relative to the Earth's orbit, the Sun's size is exaggerated by 15, Earth and Moon by 600, and the Earth-Moon distance by 40; the inclination of the Moon's orbit is exaggerated from 5° to 10°.

2017 February 11: penumbral eclipse of the Moon

The Moon arrives at its Full position as much as 19 hours before ascending node; therefore it passes through Earth's southern outer shadow. At mid eclipse the Moon is

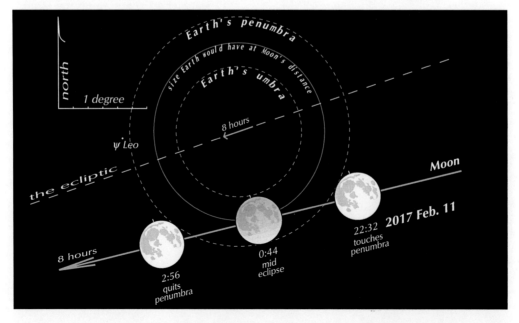

Timetables of the two eclipse seasons of 2017
All times are in Universal Time (UT).

Feb 6 14:15—Moon at perigee.

Penumbral eclipse of the Moon. February (10-)11
Number 59 of the 71 in lunar saros series 114 (971 to 2233 AD).
Feb 10 22:32—penumbral eclipse begins: first contact of Moon with Earth's shadow.
Feb 11 00:33—Full Moon (Moon at opposition to Sun in ecliptic longitude). Moon's center is exactly south of the center of Earth's shadow, as measured perpendicularly to the ecliptic.
00:44—middle of eclipse: Moon nearest to center of Earth's shadow. The penumbral magnitude of the eclipse is 1.014; that is, the penumbra reaches across the Moon and another 0.014 of its diameter.
01:11—Moon at opposition to Sun in right ascension; its center is exactly south of the center of Earth's shadow, as measured perpendicularly to the equator.
02:56—penumbral eclipse ends: last contact of Moon with Earth's shadow.

Feb 11 19:50—Moon at ascending node.
Feb 18 21:16—Moon at apogee.
Feb 21 22—Middle of eclipse season: Sun at same longitude as Moon's ascending node.
Feb 26 6:30—Moon at descending node.

Annular eclipse of the Sun, February 26
Number 29 of the 71 in solar saros series 140 (1512 to 2774 AD).
Feb 26 12:11—partial eclipse begins: first contact of Moon's penumbral cone with Earth, at local sunrise.
13:16—annular eclipse begins: first contact of Moon's antumbra (prolongation of umbral cone) with Earth, at local sunrise.
14:39—conjunction of Moon and Sun in right ascension: Moon's center exactly north of Sun's as measured perpendicularly to Earth's equator. Center of eclipse takes place at local apparent noon, with Sun and Moon on the meridian.
14:53—greatest eclipse: axis of shadow passes nearest (gamma= -0.462 Earth-radius) south of the center of Earth. As seen from the point of greatest eclipse, the magnitude is 0.992; that is, the Moon covers this fraction of the Sun's diameter. Duration on centerline is 0m 44s.
14:58—New Moon (conjunction of Moon with Sun in ecliptic longitude): Moon's center is exactly north of Sun's as measured perpendicularly to the ecliptic.
16:31—annular eclipse ends: last contact of Moon's antumbra with Earth, at local sunset.
17:36—partial eclipse ends: last contact of penumbra with Earth, at local sunset.

Partial eclipse of the Moon, August 7
Number 61 of the 82 in lunar saros series 119 (935 to 2396 AD).
Aug 7 15:48—penumbral eclipse begins: first contact of Moon with Earth's shadow.
17:22—partial eclipse begins: first contact of Moon with Earth's umbra.
18:11—Full Moon (Moon at opposition to Sun in ecliptic longitude). Moon's center is exactly north of the center of Earth's shadow, as measured perpendicularly to the ecliptic.
18:21—**middle of eclipse**: Moon nearest to center of Earth's shadow. The umbral magnitude of the eclipse is 0.252; that is, the umbra reaches across this fraction of the Moon's diameter.
18:41—Moon at opposition to Sun in right ascension; its center is exactly north of the center of Earth's shadow, as measured perpendicularly to the equator.
19:19—partial eclipse ends: last contact of Moon with Earth's umbra.
20:53—penumbral eclipse ends: last contact of Moon with Earth's shadow.

Aug 8 10:56—Moon at descending node.
Aug 16 20—Middle of eclipse season: Sun at same longitude as Moon's ascending node.
Aug 18 13:16—Moon at perigee.
Aug 21 10:34—Moon at ascending node.

Total eclipse of the Sun, August 21
Number 22 of the 77 in solar saros series 145 (1639 to 3009 AD).
Aug 21 15:47—partial eclipse begins: first contact of Moon's penumbral cone with Earth, at local sunrise.
16:49—total eclipse begins: first contact of Moon's umbra with Earth, at local sunrise.
18:13—conjunction of Moon and Sun in right ascension: Moon's center exactly north of Sun's as measured perpendicularly to Earth's equator. Center of eclipse takes place at local apparent noon, with Sun and Moon on the meridian.
18:25—greatest eclipse: axis of shadow passes nearest (gamma= 0.437 Earth-radius) north of the center of Earth. As seen from the point of greatest eclipse, the magnitude is 1.031; that is, the Moon covers the Sun and 0.031 of a Sun-width more. Duration on centerline is 2m 45s.
18:30—New Moon (conjunction of Moon with Sun in ecliptic longitude): Moon's center is exactly north of Sun's as measured perpendicularly to the ecliptic.
20:02—total eclipse ends: last contact of Moon's umbra with Earth, at local sunset.
21:04—partial eclipse ends: last contact of penumbra with Earth, at local sunset.

Aug 30 11:32—Moon at apogee.

wholly inside the penumbra—just. (At even slighter eclipses, part or most of the Moon may be outside the penumbra.) The outer half or so of the penumbra is imperceptible. A grayness on the Moon's northern part may be noticed, by viewers in the hemisphere centered on western Africa, where the Moon is overhead at midnight.

This eclipse is slight because it is toward the end of its saros series. The next, on 2035 Feb. 22, will be similar but longer before ascending node, so the Moon will pass even farther south of the shadow's center.

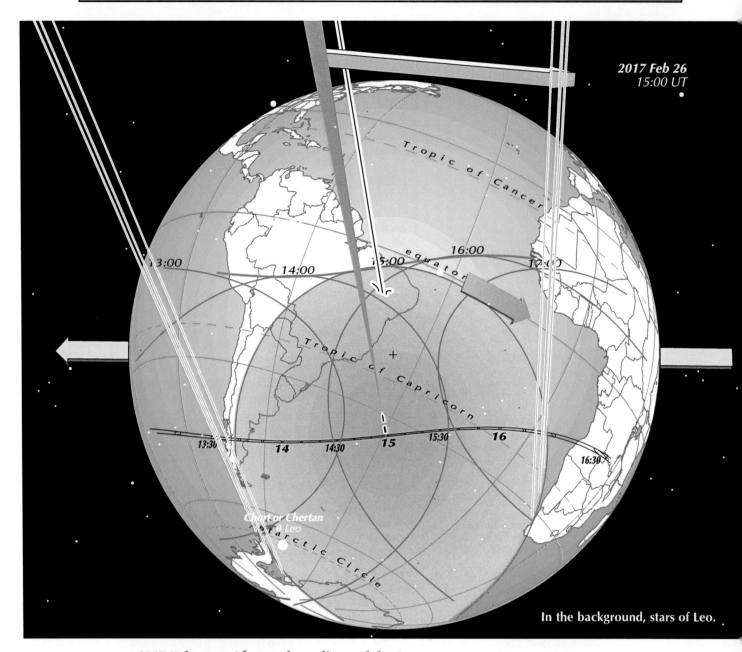

2017 Feb 26
15:00 UT

In the background, stars of Leo.

2017 February 26: annular eclipse of the Sun

The Moon circles around in front of Earth in their journey, and two weeks later descends through the ecliptic plane 8½ hours before passing between Earth and Sun. So its shadow passes across Earth's southern half. Yet the eclipse track slants generally northward across the map; this is because Earth is some way into the part of its orbit where its north pole is tilted backward.

The Moon at this time is 59.32 Earth-radii away, slightly nearer than the average of 60.3 (these are the distances between centers). Yet this is just far enough that the Moon appears very slightly smaller than the Sun—a good illustration of why annular eclipses are commoner than total ones. The ratio of their sizes is 0.992; so the brilliant ring that remains when the Moon is centered is exquisitely narrow: only 0.004 of the Sun's width.

The penumbra is never quite entirely on the Earth; at mid eclipse its southernmost edge is just off the horizon of Antactica. The antumbra, the imaginary cone from within which the Sun can be seen all around the Moon, passes across the tip of South America, and its footprint becomes narrower as it mounts the Moonward bulge of Earth in the middle of the South Atlantic, so that at the middle of the eclipse it is at its narrowest, the fit of Moon-disk to Sun-disk tightest, because the tip of the umbra is passing by at its nearest, only about 1/3 of an Earth-radius out in space. Yet the duration of annularity is longest (44 seconds) because Earth's surface is here traveling fastest in the same direction as the Moon.

This eclipse is before the middle of a saros series progressing northward across the Earth, and gradually changing from partial (the first 8 eclipses) to total (the next 11) to annular (the next 30, of which this is the 5th) to non-central annular (2—the axis of the shadow missing Earth) and back to partial (the final 16). The previous annular eclipses in this series, 1999 Feb. 16, crossed the southern Indian Ocean and then western and northern Australia; the next, 2035 March 9, will miss Tasmania but then cross the adjacent ends of the two islands of New Zealand, including at least the northern part of Wellington.

2017 August 7: partial eclipse of the Moon

The year's second eclipse season begins when the Moon, coming to its Full position, begins to pass just southward enough that it hits Earth's shadow. Full Moon is 16¾ hours before descending node, less than the 19-hour difference on Feb. 11, so the Moon goes somewhat deeper, crosses the faint penumbra; at 17:22 comes the exciting juncture—noticeable perhaps a minute or two later—when its southeastern edge is touched by the truly dark umbra. The northern curve of the umbra is on the southern curve of the Moon for nearly two hours, reaching at deepest half way to the Moon's center.

Lands around the Indian Ocean see all of this eclipse, high in the sky; Japan and eastern Australasia lose the end of it as the Moon sets at dawn; Europe sees only the last penumbral stages as the Moon rises at sunset

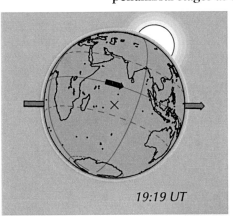 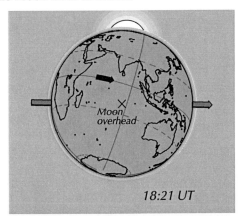 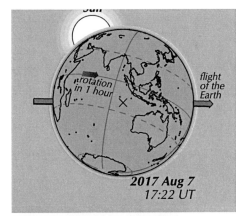

View from Moon to Earth

19:19 UT   18:21 UT   *2017 Aug 7* 17:22 UT

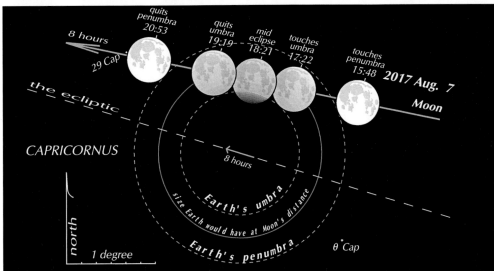

View from Earth to Moon.

A cross-America total eclipse: 2017 August 21

This is the first total solar eclipse to touch the contiguous United States since that of 1979 February 26 (whose path lay mostly in Canada). and the first to touch *only* what is now U.S. soil since 1257 June 13 (path from California via Kansas to North Carolina).

The eclipse favors us with several kinds of luck.

It is about half way to being central. That is, the Moon's passing shadow points about half way between Earth's center and northern edge. (The quantity called gamma is close to 0.5, which means that the axis of the shadow gets as near as half of an Earth-radius north of the center.) This is caused by the relatively short time (8 hours) between ascending node and New Moon. If the passage were very central (node simultaneous with New Moon, gamma near 0), as at the great Mexican eclipse of 1991 July 11, maximum eclipse would be overhead, in the tropics. If it were very un-central (longer time between node and New Moon, gamma approaching 1), it would graze the Arctic and be seen on the horizon. As it is, it is seen from our middling latitudes, at a comfortable altitude of around 60°.

Secondly, the Moon is fairly near. Again, this is not an extreme: the distance (center-to-center) is 58.3 Earth-radii, which is around halfway between average and nearest. It is near enough for the Moon to appear wider (by 3 percent) than the Sun, so that its umbra reaches Earth's surface: total, not annular, eclipse, with a maximum duration of 2¾ minutes, though if the Moon were nearer the duration would be longer.

Then there is the timing relative to Earth's rotation: the Moon passes at about 18 hours Universal (Greenwich) Time. This means: when the Sun is at the meridian for longitude about 90° west, the mid-longitude of North America. (18 U.T. is 12 noon for the Eastern time zone, though in summer it is 1 P.M. according to the distorted "Daylight-Saving Time.")

So, we've got the total eclipse happening over both the latitude and the longitudes of the contiguous states of the U.S.A. Finally there's the other part of the timing: August. This determines the track's shape; and (we hope) the weather.

Since the Moon has just passed its ascending node, its shadow travels slightly northward across Earth—but this angle is relative to the ecliptic plane. In August, Earth is moving along with its north pole tilted forward (to be most forward a month later at the September equinox), so in relation to our geography the shadow's course appears to be gently southward. The penumbra sweeps mostly over the northern hemisphere but toward the end ventures into Amazonian South America. The track of the umbra begins at latitude about 40° in the northern Pacific, and ends at latitude about 12° in the tropical Atlantic. This is the route that takes it diagonally across the U.S.A.

Finally, weather. August is, in general, sunny across the U.S., and freer of afternoon thunderstorms than July. Average cloudiness is lower in the west, lowest on the east-facing slopes of the Rocky Mountains and in the basins within them, in Oregon, Idaho, and Wyoming; and in the rain-shadow of the Rockies, that is, the upper Great Plains. Cloudiness is highest—but still not very high—in the Appalachian mountains of the east.

The Astronomical League will hold its annual convention in Casper, Wyoming, Aug. 15-19 (https://astrocon2017.astroleague.org), because of the low light pollution of the region, the dark skies around New Moon, and the likely good weather for the eclipse.

Partial phases

At 15:47 UT (8:47 AM Pacific "Daylight-Saving" Time) the outer edge of the Moon's partial shadow meets the Earth's advancing front, about 1500 miles north of the Hawaiian islands.

This penumbra spreads rapidly east over the Pacific. Soon after 16 UT, watchers (using safe methods; see page 59) in northern California should notice the first dent in the Sun's upper-right edge. The penumbra sweeps more than half of the Earth's daylight side. It covers all of North America, including Alaska, Greenland, and Panama, not all at once, though at about 18:05 all of the mainland is inside the shadow.

Places north of the centerline see the Moon's dent taken out of the southern side of

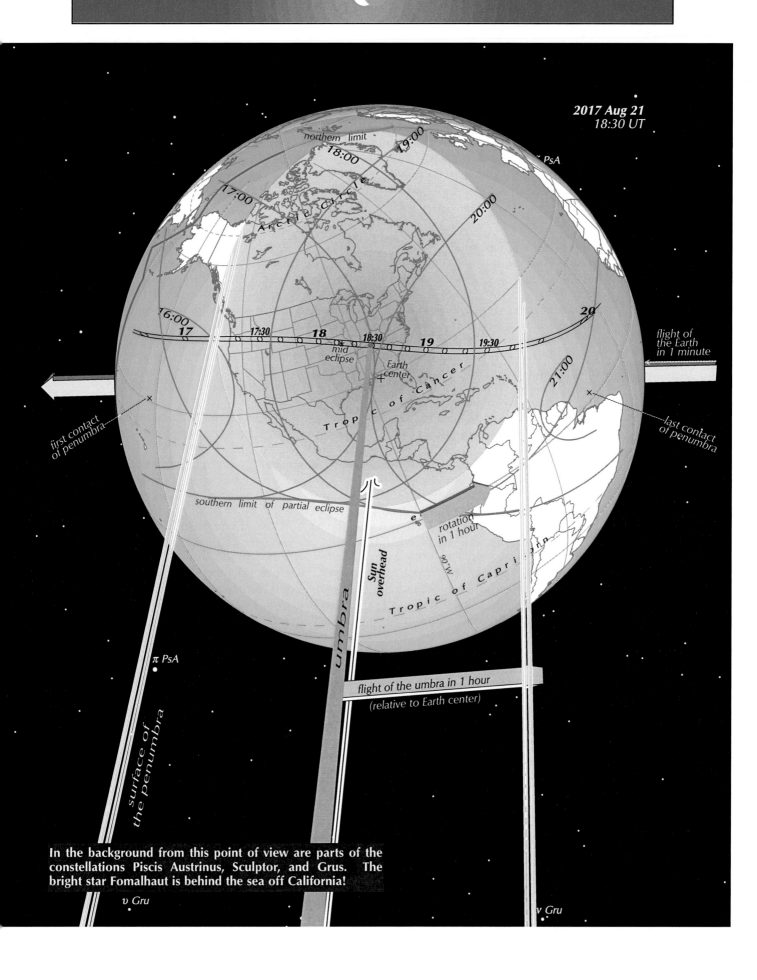

2017 Aug 21
18:30 UT

northern limit

18:00 — 19:00

17:00

Arctic Circle

20:00

16:00

17 — 17:30 — **18** — 18:30 — **19** — 19:30 — **20**

mid eclipse

Earth center

Tropic of Cancer

flight of the Earth in 1 minute

first contact of penumbra

last contact of penumbra

21:00

southern limit of partial eclipse

rotation in 1 hour

Sun overhead

Tropic of Capricorn

umbra

π PsA

ν PsA

surface of the penumbra

flight of the umbra in 1 hour
(relative to Earth center)

In the background from this point of view are parts of the constellations Piscis Austrinus, Sculptor, and Grus. The bright star Fomalhaut is behind the sea off California!

υ Gru

υ Gru

the Sun, and vice versa. Places near the centerline see the deepest and longest partial eclipse, places at or near the limits may see no appreciable eclipse. Places to the west of a more or less north-south line through the Mississippi valley see the eclipse in the morning; east of that, the eclipse is seen as the Sun goes down the afternoon sky.

The penumbra comes close to missing Earth on the north, so the line denoting "northern limit" of partial eclipse is short. It is a curve in the sea off northeastern Greenland. Notice that the north pole is *inside* this line, so that "northern limit" is something of a misnomer, though there is no substitute for it. At this northern-summer time of year, a sailor south of the north pole in the direction of Russia sees the Sun move *over* the north pole, and at about 18 UT sees it very slightly eclipsed.

Story of the total eclipse

An hour after the penumbra touches down, the umbra or central cone of the shadow does so, farther north and also farther west, because the Earth has been rotating for an hour.

The umbra brushes—most swiftly at first—for 28 minutes across open ocean, then over solid America for 91 minutes, from 17:16 UT (10:16 Pacific DT) to 18:47 (14:47 EDT). The path lies across the middles of Oregon, Idaho, Wyoming, and Nebraska; a corner of Kansas and a small corner of Iowa; Missouri, Illinois, Kentucky, Tennessee; the meeting of Georgia with North Carolina; and down the middle of South Carolina. Then 75 minutes over the Atlantic, till 20.02 UT, when the umbra quits the Earth at sunset, about 600 kilometers off the west coast of Africa.

Maximum eclipse happens near what I think of as the geographical focus of the U.S.A., where the Mississippi and Ohio rivers meet, and where Illinois meets Kentucky. Actually it depends on which kind of maximum is meant. "Greatest eclipse," when the axis of the shadow passes closest to the center of the Earth (gamma at a minimum of 0.44 of Earth's radius), is at 18:25.5 UT, and in Kentucky, near a small town called Hopkinsville. But the moment of greatest duration, affected slightly by other factors, occurs earlier, about 18:21, near Carbondale, Illinois.

(And the moments of conjunction, when the Moon is exactly north of the Sun in right ascension, and New Moon, when it is north of the Sun in ecliptic longitude, are different again: 18:13 and 18:30 UT.)

Since mid eclipse happens as the Sun crosses the meridian, the local solar time is 12. Illinois is in the eastern side of the Central time zone, and fortunately the west end of Kentucky is too, so for both of them 18:25 UT is clock time 13:25. It would be 12:25, but for the one-hour political distortion of clocks in summer.

The duration of totality on the centerline is 1 minute 59 seconds on the west coast; 2m 40s at the maximum (less than half the greatest possible, 7m 31s); 2m 39s on the east coast.

The altitude of the eclipsed Sun, as seen from the centerline, is 38° at the west coast, 64° at the maximum, 61° at the east coast.

The width of the path is 99 kilometers at the west coast, 115 at the maximum, barely less at the east coast.

The speed of the umbra over the ground reaches, at maximum eclipse, a minimum of 0.65 kilometer per second, because this part of Earth's surface is nearest to the Moon and moving in the same direction.

(Some of these figures differ slightly between calculators, such as the U.S. Naval Observatory, which makes the maximum duration about 4 seconds longer, and experts Fred Espenak and Jay Anderson.)

Some places inside the path are Casper, Wyoming, on the centerline; Idaho Falls; North Platte, Nebraska; in Missouri, Kansas City on the southern edge and St. Louis on the northern; Nashville, Tennessee; in South Carolina, Greenville and Columbia not far north of the centerline, Spartanburg on the northern edge, Charleston south of the centerline.

From cities not far outside the path, partial eclipse will be tantalizingly deep: on the north, Portland, Bend, Omaha, Cincinnati, Knoxville; south of the path, Eugene, Boise, Denver, Topeka, Chattanooga, Atlanta.

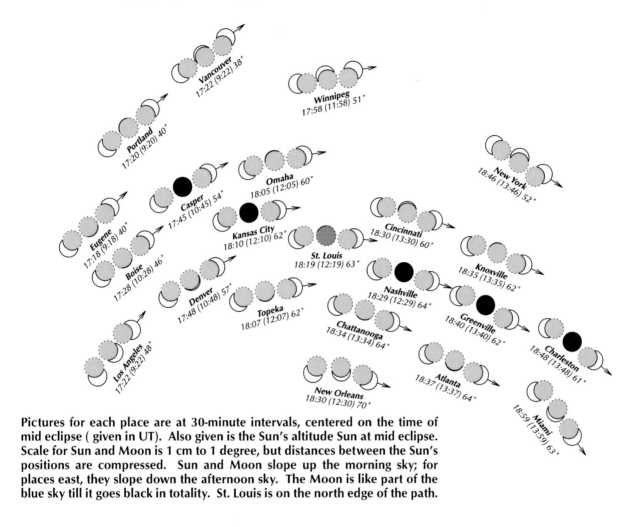

Pictures for each place are at 30-minute intervals, centered on the time of mid eclipse (given in UT). Also given is the Sun's altitude Sun at mid eclipse. Scale for Sun and Moon is 1 cm to 1 degree, but distances between the Sun's positions are compressed. Sun and Moon slope up the morning sky; for places east, they slope down the afternoon sky. The Moon is like part of the blue sky till it goes black in totality. St. Louis is on the north edge of the path.

Edge versus centerline. Observers customarily try to position themselves on or near the centerline, so as to have the longest view of totality and the corona and its main feature, the corona.

An alternative philosophy, advocated by the late Tom Van Flandern, is to be just three or four miles inside one of the edges—preferably the southern, since the southern limb of the Moon is rougher with mountains. For then the "edge-effects" just before and after totality stretch out in slow motion. These are the shadow-bands, Baily's beads, the Diamond Ring, and the red chromosphere.

The length of totality is of course reduced, but much less than you might think, since it follows a curve: 20% of the way from center to edge it is reduced only to 98%; 90% of the way, to 50%; 95% of the way, to about 33% (and this is about where the edge effects are at maximum).

On 1991 July 11 I found this to work dramatically well. I watched shadow-bands unmistakably for the first time (though they were even better in 1999 on the centerline); the Diamond Ring was overpowering; and the chromosphere, which I had never before glimpsed, lingered throughout totality, sidling around the southern edge!

Yet such long eclipses are likely to show the *least* enhancement at the edges, because the Moon's circle is so much larger than the Sun's. At shorter eclipses such as 2017 Aug. 21, where the two circles fit tighter together, there could be wider edge-zones extending deeper into the path, where a feast of beading and other edge effects is seen.

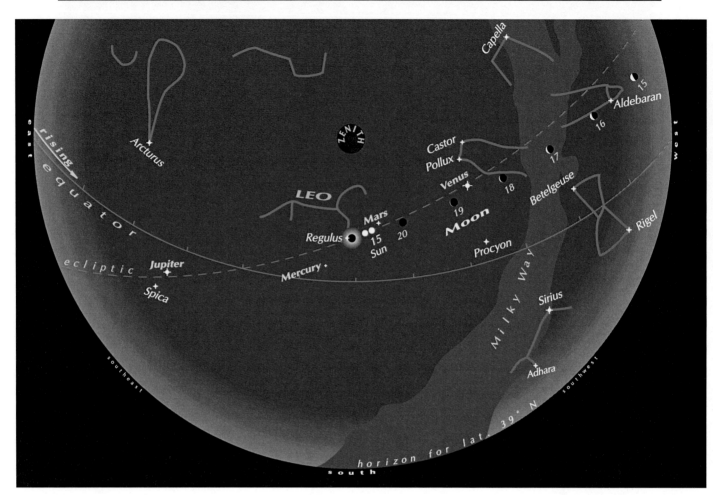

The dome of the sky, as seen from Illinois/Kentucky at 18:25 Universal Time. The Moon and Sun are exaggerated 4 times in size. They are also shown over the preceding week at the same clock time (the Sun at 2-day intervals) to suggest how the Moon, moving 13 times as fast, arrives at the ecliptic in time to cover the Sun.

The Milky Way and some constellation figures are shown to indicate the general situation of the sky, though they are far below visibility. Nor is it likely that all the stars and planets marked will be seen.

The Sun is at its highest altitude, 64°. Seen from the west coast 68 minutes earlier, the eclipsed Sun was half way to the left, 38° above the eastern horizon; Jupiter was just about to rise. Seen from the east coast 22 minutes after the time of the picture, the eclipsed Sun will still be high, 51° above the southwestern horizon; Sirius and Betelgeuse will be almost down to the horizon.

The sky revealed during totality

Even though this is not the longest of eclipses, it may be worth while taking some seconds to look away around the sky, for the rare experience of "stars by daytime." They happen to be an interesting set.

If conditions allow any at all to appear, the first will be, as usual, Venus, probably before totality begins.

Of the five bright planets, only Saturn is below the horizon (being 112° east of the Sun). The other four are arranged almost symmetrically: to the east (left), Mercury and Jupiter (10.5° and 51° away), on the west Mars and Venus (8° and 34°). So these four are fanned over a span almost 90° wide.

The "royal" star Regulus is barely over a degree to the east of the Sun. Your eye, or more likely your camera, may find it nestled within the streams of the corona.

Nine first-magnitude stars (out of about 16 ever visible from this latitude) are above the horizon, though certainly not all will be seen. The order or brightness is: Venus (magnitude –4.0), Jupiter (–1.8), Sirius (–1.5), Arcturus (0.0), Vega (0.0, barely risen on the northeast horizon), Capella (0.1), Procyon (0.4), Spica (1.0), Pollux (1.1), Regulus (1.4), Castor (1.6), Mars (1.8), Mercury (3.3).

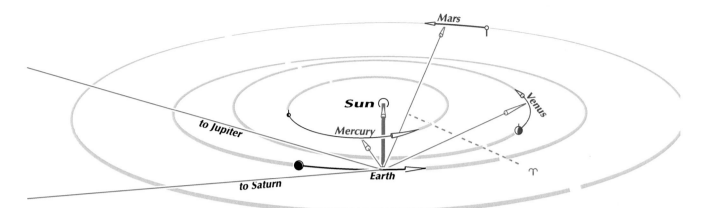

Spatial diagram showing why the planets appear where they do. Curved arrows are their courses in August. Straight arrows are sightlines from Earth to each planet and (thick) to the Moon and Sun, at the time of the eclipse. The dashed line and ram's-horns symbol indicate the vernal equinox direction (longitude 0). The view is from ecliptic longitude 329°, latitude 15°

Mercury is approaching its inferior conjunction (Aug. 26). Venus is not far past its westernmost elongation (June 3). Mars has appeared in the morning sky after being behind the Sun on July 27. Jupiter will be behind the Sun on Oct. 26, and Venus will be at conjunction with it on Nov. 13. Saturn is far westward in the evening sky, having been at opposition on June 15.

DO NOT LOOK DIRECTLY TOWARD THE SUN, <u>except when it is totally eclipsed</u>. Some safe ways during partial eclipse are:

—<u>Projected</u> image through a pinhole, or through a telescope, onto a sheet of paper.

—Looking through <u>No. 14 welder's glass</u> (or No. 12 if the Sun is dim).

—Eclipse-viewing spectacles if they are from a provider you know you can trust.

Do NOT use smoked glass, exposed film, crossed polarizing filters, colored water. They may seem to dim the Sun, but infrared rays get through.

Do NOT look through a telescope unless you really know what you are doing. Any filter should be over the front, NOT behind the eyepiece.

When the Sun is dim (low or through cloud) still take only BRIEF naked-eye looks. Retinal damage can happen without hurting.

There is nothing dangerous about sunlight during total eclipse. It's just that we normally have an instinct not to look at the Sun; during an eclipse we are interested in doing so. During totality, it's safe (and gorgeous) to look!

Saros history

This eclipse is the 22nd in solar saros series 145: 77 similar eclipses, from 1639 Jan. 4 to 3009 Apr. 17, progressing southward across the Earth, and gradually changing from partial (the first 14 eclipses) to annular (one eclipse, 1891 June 6) to annular-total (one eclipse, 1909 June 17) to total (41, of which this is the 5th) and back to partial (the final 20).

The previous eclipse in the series, 18.03 years earlier and a third of the way around the world, was 1999 Aug. 11, when the totally eclipsed Sun was seen by probably more people than ever before. The track crossed the southwestern corner of England (I originally intended to see it from Land's End), many European cities such as Strasbourg, Munich, and Bucharest; then Turkey, where I took a party of people to see it, on a hill above a village called Shenyurt, near a town called Turhal. The track of totality went on across Iran, the coast of Pakistan, and India.

The next in the series, 2035 Sep. 2, will start in the morning across the three northern provinces that China has absorbed from non-Chinese peoples: Xinjiang, Inner Mongolia, and Manchuria; then cross North Korea, and Japan just north of Tokyo.

The saros

Eclipse-seasons run in waves through the year, but within this pattern there seems at first a jumble of eclipses of all kinds. The more you contemplate the eclipses in a year, or neighboring years, the more symmetries and balances you find. But one long-range rule was noticed long ago:

For any eclipse (solar or lunar), the two eclipses most like it are those which happen about 18 years before and after.

This, like other patterns in astronomy, is reminiscent of the patterns of music. A tone most resembles its octave. There are many tones in between, but it *least* resembles— is most discordant with—its closest neighbors. And the eclipse calls to its cousin 18 years away; it contrasts with, rather than resembling, the other eclipses of its own and neighboring years.

But solar years are not the most appropriate units in which to measure eclipses. There is no reason why eclipses should occur any integral number of years apart. Since all eclipses happen at New or Full Moon, the accurate unit to use is the synodic month: the period of revolution of the Moon around the Earth, from New to New or from Full to Full. The synodic month is 29.53 days. (This is an average; the actual time from one New Moon to the next varies quite a bit, and the average itself slowly lengthens over the centuries.) Solar eclipses must be integral numbers of synodic months from all other solar eclipses, and lunar eclipses likewise from all other lunar eclipses.

Well, in synodic months this interval between most-similar eclipses turns out to be 223. In days that translates to 6585.32. In years it is about 18.03; that is, it is 18 years plus 10.32, 11.32, or occasionally 12.32 days, depending on how many leap-days are inserted in this particular span of calendar years. In these latter numbers, the difference between the 10, 11, and 12 has little importance: it is merely a result of our artificial calendar with its years of two different lengths. The fractional part, 0.32, is important, since it represents a difference of nearly a third of a day between otherwise similar eclipses.

Knowledge of this interval of about 18 years gives a rudimentary power to predict eclipses. Given the total solar eclipse of 1981 July 31, which ran through Siberia, you know that there will be an eclipse 6585.32 days later. From 1981 July 31 to 1999 July 31 is $18 \times 365 = 6570$ days, $+ 4$ leap-days (in 1984, 1988, 1992, 1996) $= 6574$ days; so 11 more days are needed to reach the total of 6585; so the eclipse will be on 1999 August 11. It will then fall 0.32 of a day later. In that third of a day, the Earth will have made an extra third of a turn eastward, so the eclipse will be seen a third of the globe backward (westward): the eclipse will run through Europe.

This third-of-a-day difference considerably spoils the predictive effect. The people of early civilizations lived in worlds confined to small patches of our world. Having seen a solar eclipse they would probably *not* see and not hear of the one 18 years later, happening a third of the way around the globe. At any rate, they would not be in the track of totality twice, though it must be remembered that partial eclipse can be seen over a far wider area: it is more conceivable that a dark near-eclipse in the morning could be connected with another in an evening 18 years later. After three of these periods (54 years and 1 month) the eclipse returns to nearly the same face of the Earth, so that it could be seen at about the same time of day and height in the sky. But the exact central track would be some hundreds of miles away, so that, one of the two times at least, it would be only a partial eclipse that was seen. Then there are the lunar eclipses, which can be seen from all over one half of the globe, so that it is easier to catch more of them. (Early people were more often out under an unobstructed night sky, especially in the hours before dawn.) One could, for example, without moving from America, catch the shadow moving off the Moon as it rises at the beginning of the night of 1992 Dec. 9; then see this shadow again on the Moon midway through the night of 2010 Dec. 21. Another possibility, to which we shall return when the "Bead" diagram has made things clearer, is that it was noticed that *solar* eclipses are regularly followed by *lunar* eclipses over about the same region only 9 years later.

The Babylonians began to keep lists of eclipses at least as far back as 747 B.C. (Actually such a clerical habit could well have been inspired by the Assyrians, who

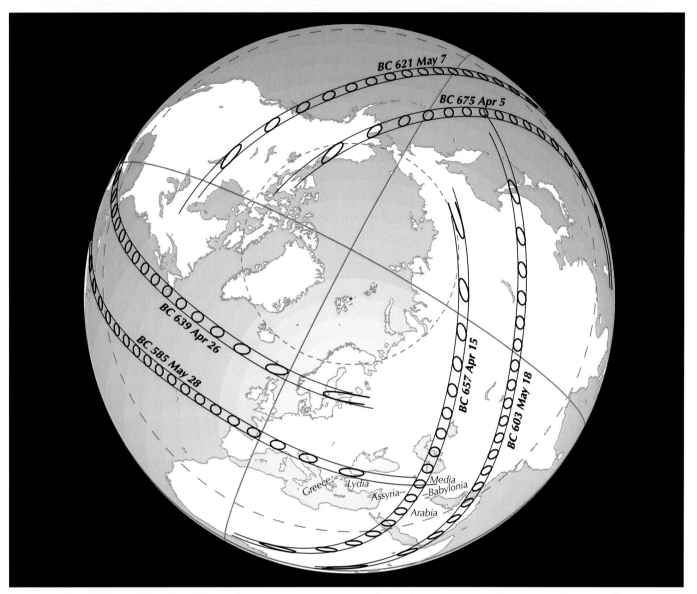

Paths of some total solar eclipses that were successive members of a saros series. (We count them as eclipses 28 to 33 in series 57, which ran from 1162 B.C. to A.D. 137.) According to the Greek historian Herodotus, the sage Thales predicted an eclipse, presumably that of 585 B.C., which stopped a battle between the Medes and Lydians. Could Thales have learned from Middle Eastern people about the eclipse of 603 or that of 657?

owed a lot of their imperial success to efficient office practices. The date falls in the short period of Assyrian weakness before the accession of Tiglath-Pileser III in 745 B.C. and the rise of the Late Assyrian empire.) After keeping lists for perhaps 150 years, the Babylonians noticed the 18- or 54-year recurrences. The first evidence we have is that they recorded their *failure* to see an eclipse they had predicted for 568 B.C., in the time of Nebuchadnezzar. (It happened, but could not be seen from Babylonia.)

In what we loosely call "Babylonia" (the southern part of Mesopotamia or Iraq) two peoples lived intermingled: we call them Sumerians and Akkadians. Sumerian, the world's oldest written language, vanished and has no certain relatives; Akkadian, giving rise to the later Babylonian and Assyrian, was a language of the Semitic family (like Arabic, Hebrew, and Ethiopic). The 60-based numbering system, which has given us hours and degrees divided by 60 and 60 times 60, was presumably an invention of the Sumerians. They had a word for "3600": *sharu* (which can also be written in its shortened form *shar*). The list of the Sumerian kings, composed about 2000 B.C., assigned to the ten kings "before the Flood" (that is, before about 4000 B.C.) reigns of fantastic length, measured in many *sharu* of years. The word was borrowed into Akkadian, where it gained a few other usages. Berossus, a late Babylonian living about 300 B.C., wrote a version of these histories in Greek, transcribing *sharu* as *saros*. Later rationalists interpreted *saros* in the king-lists as 3600 days (or about ten years). Around A.D. 1000 a Greek lexicon-encyclopaedia known as the *The Suda* was compiled by a writer who has come to be called "Suidas"; he by some misunderstanding took *saros* to mean 18½ years. Edmund Halley knew of the unit of 223 synodic months: he made a set of observations of the Moon over that period, which he published in 1710 as an appendix to the second edition of Street's *Caroline Tables*. He applied "saros" to this eclipse-period. By the nineteenth century, astronomers were assuming that it had been thus applied by the Babylonians (or "Chaldaeans"). This seems not quite true: the Babylonians had the word, also must have been aware of the period in some form in order to predict eclipses, but it was apparently not they who put the two together.

What underlies the saros rule? Why are there eclipses 223 synodic months apart (rather than, say, 222, or 225), and why are they similar?

223 happens to be a prime number. That, however, is not the reason - just a pleasing extra fact!

Consider these units of time which affect the Moon and therefore eclipses. If time-units seem boring, think of them as *rhythms*, for that is what they are:

—The primary one is the synodic month of (average) 29.53059 days, since every eclipse must be at a New or a Full Moon.

—The time-unit that describes the rhythm of the rising-and-falling component in the Moon's motion, above and below the ecliptic plane, is the nodical month: the period between the Moon's crossings of one of its orbit's nodes. Its average length is only 27.21222 days. It is also called the draconic month, because of the eclipse *dragon* that swallows the Sun. For this is the time-unit that governs whether, at a New or Full Moon, an eclipse happens or not: eclipses can happen only near a node. How near, determines how central the eclipse is, that is, how near the Moon passes to the center of the Earth's shadow, or how near the Moon's shadow passes to the center of the Earth. A more central eclipse is of a more important kind: total versus partial or penumbral; and more central total eclipses are longer, higher in the sky, seen in easier parts of the Earth, than less central ones. Also, whether the node is ascending or descending has interesting though less vital effects on the eclipse: the path of the shadow curling northward or southward across the Earth's geography, or the path of the eclipsed Moon slanting northward or southward among the stars.

—The time-unit that describes the outward-and-inward rhythm in the Moon's motion, its varying distance from Earth, is the anomalistic month, or period between the Moon's perigees (moments when it is nearest to the Earth). The Moon's distance makes the difference between annular and total eclipses; in lunar eclipses, a nearer Moon appears larger. The average length of the anomalistic month is 27.55455 days.

—The year, the period in which the whole Earth-Moon system goes around the Sun, determines some other characteristics of an eclipse. Eclipses in, say, March are seen out in the same direction is space: they have the same stars in the background. The season also affects the geography of the eclipse track: the Earth travels with its north pole in March leaning backward, in June leaning toward the Sun, in September leaning forward, and in December leaning away from the Sun, and this causes the eclipse track to slope and curve in corresponding ways across the globe.

—Finally, the day. The time of day is equivalent to the direction the Earth is facing; and that means the longitude over which the eclipse happens. Is longitude important? Certainly it is: it is the difference between Europe and America, between America and Asia.

The saros is amazingly close to being a common multiple of these units. It is 223 synodic months, and this turns out to be very close to an integral number of nodical months, and another integral number of anomalistic months, as well as fairly close to 18 years.

```
synodic months:              223 ×  29.53059  = 6585.3216 days
nodical (draconic) months:   242 ×  27.21222  = 6585.3572 days
anomalistic months:          239 ×  27.55455  = 6585.5375 days
(years:                       18 × 365.2425   = 6574.365  days)
```

This means that at the 223rd New Moon after any given eclipse, the Moon is again at the same node, so there is an eclipse. And, since a whole number of anomalistic months has passed, the Moon is at the same distance. And the time of year is the same (plus only about 11 days).

The only large difference is in the time of day, that is, the position of the Earth in its rotation. 6585.3216 is 10.9566 longer than 6574.365; that is, the saros is almost exactly 11 days longer than the length of 18 true years. However, that length is a fractional number of days; any particular span of 18 calendar years is either 6573, 6574, or 6575 days, so the difference is 12.3216, 11.3216, or 10.3216 days. If the saros were an integral number of days—6585—the successive eclipses would be seen over exactly the same part of the Earth; but it is not—it is 6585 and a fraction—because the rotation of the Earth is not locked to the cycles of the Moon; and so at each eclipse the Earth has rotated another fraction. This results in the greatest dissimilarity between two eclipses a saros apart. If one of them is centered over, say, Europe, the next is centered over America.

There is a simpler way of expressing the saros: in terms of eclipse seasons and eclipse years. 19 eclipse years are roughly the same as 18 years. The eclipse year is really a derivative of the nodical month: it is because the nodes of the Moon's orbit move backward (because the nodical month is shorter than the synodic) that eclipse seasons shift earlier in the year. This means that after a year when eclipses are happening in, say, June, 9 years later they are again happening in June (the other eclipse season having reached this time of year), and another 9 years later they are again happening in June: but then it is the *same* eclipse season that is falling in June, so that the eclipses are *similar* ones. It is not really so surprising that after this cycle of years the various kinds of month—other rhythms in the Moon's motion—are again approximately in phase. The Moon's motion is influenced by many, many factors, such as the positions of other planets, but most important are its position in relation to the Earth (with its slight equatorial bulge and its seasonal tilt) and the Sun. When after 18 years eclipses are happening around the same position in space, with the Moon passing an Earth in a similar attitude at a similar distance from the Sun, we would expect the eclipses to be similar.

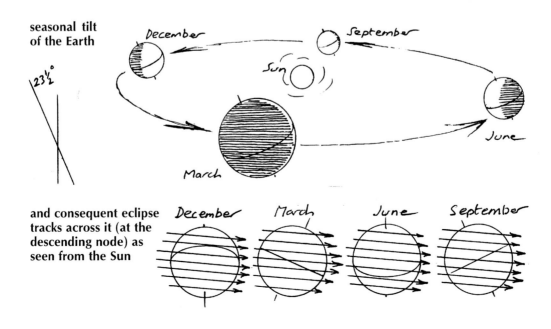

seasonal tilt of the Earth

and consequent eclipse tracks across it (at the descending node) as seen from the Sun

Saros series and their evolution

If an eclipse's next of kin are the eclipses at the time-interval of one saros before and a saros after, then those have other similar eclipses another saros away. When you pick up any eclipse, you are picking up a string: a saros *series*. All eclipses (solar and lunar) belong to such series, of which about 82 run concurrently (this being the number of eclipses in 18 years).

Are these series infinite in length? They would be, if the coincidences between the various time-units were exact. Let us suppose that 223 synodic months and 242 nodical months and 239 anomalistic months all came out to exactly 6585 days—no, let's make it 6574, the closest we can get to 18 years. Then a series of, say, total solar eclipses at the Moon's ascending node, with the Moon at a certain distance, would stretch infinitely into the past and future. They would keep occurring on, say, May 16 (or 15 after a leap-day), at the same time of day over the same part of the Earth; the Moon each time would be at the same distance; the duration of totality and other circumstances would be the same. There would in fact be only four saros series, in May and November, a lunar and a solar series at each node (or, if the seasons were well centered, six series: two marginal lunar eclipses and one central solar one, or vice versa, near each node).

But the coincidences are not exact, and so the eclipses of a saros series do not remain identical.

Because the saros length ends in a fraction of a day, the eclipses keep happening over different parts of the Earth.

Because the saros is 11 days longer than the average year, successive eclipses of a series fall 18 years and 11 days later. After about 3 eclipses (54 years) they have advanced a month; after 6 eclipses (a century) they are happening two months later; after about 33 eclipses (six centuries) they return to the original time of year.

Because 239 anomalistic months is a few hours longer than the exact saros (that is, than 223 synodic months), the Moon's distance at each eclipse in a series gradually changes. So a series of solar eclipses that starts with the Moon rather distant, and the eclipses annular, can evolve so that later eclipses in the series are annular-total, and still later ones total. Another series may evolve in the opposite direction, becoming gradually more distant: total to annular-total to annular.

Actually, the factors mentioned so far bring change but not real evolution: they are periodic, rather than secular. Eclipses of a series return around the Earth each third time to America; they return around the year each thirty-third time to July; they can even cover a reversal in the Moon's distance, so that they return to being annular or total. But:

Because 242 nodical months is about 50 minutes longer than the saros, the Moon at each eclipse reaches the node a little later. This change—caused by the smallest of the differences between the synodic month and other units—is the most profound. And that stands to reason: "nodality" (to invent a word for "nearness to a node"), together with "syzygality" (nearness to New or Full Moon), is what causes there to be an eclipse at all. The other factors affect important qualities of the eclipse, but not its existence.

Suppose that one time the Moon crosses the descending node exactly at New Moon. The eclipse is maximal: the axis of the Moon's shadow crosses the center of the Earth. But next time it crosses later, hence passes a few hundred miles farther north. Of course, this also implies that there were earlier eclipses passing farther south. At the central eclipse of any real series, the Moon will not cross the node *exactly* at New Moon nor the track cross the *exact* center of the Earth; 1991 July 11 is remarkable in that the difference in time is as small as 3 minutes and the quantity called *gamma*—the measure of centrality—as small as 16 miles.

One can see that the paths of successive eclipses gradually grind across the latitudes of the Earth, from southern to northern. But that cannot be all: if the saros series has a center, it must have a beginning and an end. Toward the end, the tracks move to the Arctic. They become shorter, because crossing smaller chords of the globe; it is harder for an observer to get to them, and having got there he sees the eclipse low in the sky. Then the track of the umbra moves off the Earth entirely. The broad penumbra still

Successive eclipses of a series at the descending node (each 7th shown):

—at the ascending node:

touches, so partial eclipses are seen. Their tracks too shrink; less of the Sun is seen covered for briefer times. There comes a last and slightest eclipse in the series. At the New Moon that comes 18.03 years after this, the whole shadow just misses the Earth's northern rim; the Moon descends to its node too late, too far past its passing of the Earth.

Similarly at the beginning. There must have been a time when the Moon descended through the node just too soon, so that as it then went past the Earth (at New Moon) its shadow dipped just too far south. One saros later, it came fifty minutes later to the node; the northern edge of the shadow touched, and there was a barely discernible partial eclipse in Antarctica. At each successive saros-period, the penumbra gained broader footholds on the southern oceans, until the umbra first touched and the annular or total eclipses began.

The story is the same for a series of eclipses taking place at the ascending node, except that then the successive paths pass farther south. This rule can be hard to remember—eclipses at the ascending node make north-sloping tracks across the globe but each track is successively farther *southward*—unless one bears in mind what it is based on, namely that each track is slightly *onward* in relation to the node. The first eclipse of an ascending-node series is achieved when the Moon climbs through the node late enough that the bottom of its shadow then brushes the north of the Earth; as it keeps reaching the node later, the interval between node and New Moon decreases and the shadow eats southward.

There is another consequence: an eclipse at the *beginning* of a series happens at the *end* of the eclipse season in which it falls. For the Moon is still reaching its syzygy (New or Full Moon position) rather too long after crossing the node. Vice versa with eclipses late in their series. Or we could put the thing the other way around: the first eclipse in any season belongs to an old series, the last eclipse in a season is an early member of a young series. The center of an eclipse season is when the line of nodes is lined up with the Sun; if an eclipse falls here, it is at the center of its series, when node passage coincides with New or Full Moon.

Each saros series is a story, with beginning, middle, and end. It has a finite life, though a long one. The most common number of eclipses in a series is 71. But the number in solar series ranges from 69 to 82, averaging 72.1; in lunar series it ranges from

70 to as high as 84, averaging 73.8. To find the spans of time that these series cover, we can multiply the number of saroses (saroi?) from eclipse to eclipse by the saros length of 18.03 years. Thus 69 eclipses span 1226 years; 71 eclipses, 1262 years; and 84 eclipses, 1496 years.

There is a grandeur to this cycle of time, with its tolling of luminous events and its Babylonian name and its duration of far over a thousand years. Numerologists and eschatologists (as I realized when one came to ask me about it) would like to relate it to periods human and divine, the end of the world and the conversion of the Jews.

Their image of "saros" may be a crystalline span of, say, 1262 years, like the reign of some king before the Flood. But not only are the series of various lengths; their edges are soft. A series, like a hill, rises to a central summit, but like a hill the limits of its two slopes are subtle to define. All series begin and end with eclipses of the most negligible kind. The first few and last few of a lunar series are eclipses that will certainly not be seen, since only the outermost penumbra of the Earth will briefly touch the Moon. At the first few and last few of a solar series, only someone in the Arctic or Antarctic, observing carefully, could briefly see a sliver of the Sun covered.

These eclipses shrivel from the negligible into the theoretical. A small difference in method of calculation can make the difference between an eclipse and no eclipse. Or in one of the assumed numbers, such as the radius adopted for the fuzzy-atmosphered Earth, and especially the irregular Moon. This is a much-argued number called k: some use the Moon's average radius, some a smaller to allow for sunlight coming through valleys in the Moon's profile. Oppolzer's *Canon of Eclipses* gives a partial solar eclipse on 2156 March 21 that Meeus finds to be no eclipse. And some authors find several penumbral lunar eclipses where others find none.

This means that the number of members given for a saros series may vary. Indeed, it makes suspect, or at least makes less important, our statistics about the frequency of eclipses.

Yet the idea of the saros itself lends a greater interest to the otherwise negligible events. *Every* penumbral lunar or partial solar eclipse represents an opening or closing stage of some saros series. A shadow that flicks unobserved across the polar regions of the Earth or of the Moon is the precursor of thirty or more that, eighteen years apart, will inexorably move deeper and deeper. Each such eclipse is at the thin end of one of these thousand-year dark strands that, laid together eighty thick, individually ceasing but collectively persisting, compose the immense and almost eternal time-braid of the eclipses.

Numbering the saros

You know that something is formally embedded in astronomy when it has been given a numbering-system. This was done for lunations (cycles of the Moon around he sky) by the great Moon-theorist Ernest W. Brown in a publication of 1933, and for saros series by the Dutch astronomer G. van den Bergh in a publication of 1955.

Lunations are successive cycles of the Moon from New Moon to New Moon (so that they have the length of the synodic month, averaging 29.53 days). Brown gave the number 1 to the lunation that started on 1923 Jan. 16. We have to use negative numbers for the lunations of earlier years (those of 1922 are −11, −10, −9, −8, −7, −6, −5, −4, −3, −2, −1, and 0). Lunations accumulate at a rate of about 12.37 in a year (1237 in a century). So for example the first lunation to start in 1991 (on Jan. 15) was number 842; and the 1991 July 11 eclipse occurred at the New Moon that started lunation 848.

As to saros series, Van den Bergh assigned the number 1 to a lunar and a solar series that were running in the early second millennium B.C. However, these series are not so simple to number. Whereas lunations follow each other head-to-tail, saros series overlap, about 80 of them at any given time. So the next thought is that successive numbers will be given to new series as they start up. However, even this is not practicable. For whereas lunations begin and end sharply at New Moons, saros series have their soft and sometimes debatable beginnings. The most definite thing about a saros series is its peak: the eclipse at its center that passes closest to the center of the Earth. There may be as many as 40 or as few as 26 eclipses before this central one (and a similarly variable number after it); and so a series that is really earlier, in the sense that it will reach its maturity earlier, may start later. Therefore what must be done is to number the series in the order in which they reach their peaks.

They do this at a rate of one at about each 358th lunation. The interval of 358 lunations (which works out at nearly 10,572 days, or 29 years less about 20 days) was therefore called by Van den Bergh the inex. (It is the interval at which series go in and out of existence.) Each inex, another series reaches its central peak or climax. We can also say that new series are born, and old ones die, at a rate of one each 358 lunations, but this is an average.

For example, the very central total solar eclipse of 1991 July 11 was the peak of series 136. One inex (358 lunations) later, series 137 reaches its peak with the very central annular eclipse of 2020 June 21.

But to illustrate the varying intervals between the slight partial eclipses that begin series:

```
1917 July 19   first of series 154   lunation   -62
1928 June 17                   155              68    130 from last
2011 July 1                    156            1095   1027
2058 June 21                   157            1676    581
2069 May 20                    158            1811    135
2098 Oct. 24                   164!           2175    364
```

There are parallel sets of numbers for solar and lunar eclipses. Here are examples from solar and lunar series that happen to be called by the same numbers:

```
eclipses in        at New Moon   eclipses in        at Full Moon
solar series 136   of lunation   lunar series 136   of lunation
                                 1986 Oct. 17            789
   1991 July 11        848
                                 2004 Oct. 28           1012
   2009 July 22       1071
                                 2022 Nov.  8           1235
   2027 Aug.  2       1294
```

Thus from a solar to the next lunar eclipse of the same saros number is (for no particular reason that I know of) 164½ lunations, or from lunar to solar 58½ (these adding up, of course, to 223, the saros).

It works out that adjacent eclipses (two weeks apart) are 12 saros numbers apart (if the order is solar-lunar) or 26 (if the order is lunar-solar). Thus if there are three eclipses in a season, the first and third (the two solar, or two lunar) are 12 + 26 = 38 saros numbers apart. For example in the middle of 1991 the eclipses of June 27, July 11, July 26 (lunar, solar, lunar) were in series 110, 136, 148.

Odd numbers are given to all solar eclipses that happen at the ascending node of the Moon's orbit, and all lunar eclipses at the descending node. Is this not perverse?—why not make it odd for all at the ascending node, even for all at the descending? But what the rule really means is that all the eclipses of one type of eclipse season have odd numbers. For within any season the solar and lunar eclipses occur at the opposite nodes. Here are three eclipse seasons:

```
          eclipse              at node      in series
     1991 Jan. 15   SOLAR    Ascending        131
          Jan. 30   lunar    descending       143
          June 27   lunar     ascending       110
          July 11   SOLAR    Descending       136
          July 26   lunar     ascending       148
          Dec. 21   lunar    descending       115
     1992 Jan.  4   SOLAR     Ascending       141
```

After any eclipse, the next one of the same kind (the next solar after a solar or lunar eclipse after a lunar) is usually in the next eclipse season and is 6 lunations ahead (a period sometimes called a semester). It turns out to belong to a saros series which is 5 numbers later. 1990 was one of these simple years:

	New Moon	solar eclipse in series
1990	Jan. 26	121
	Feb. 25	.
	Mar. 26	.
	Apr. 25	.
	May 24	.
	June 22	.
	July 22	126

When a three-eclipse season arrives (with two solar or two lunar eclipses) it makes this look more complicated. The next eclipse after a given one can then come sooner: 5 lunations ahead or only 1, though the eclipse 6 lunations ahead still occurs. (Of course there cannot be eclipses separated by the intervals of 2, 3, or 4 lunations.) Thus in the year 2000 there is a two-eclipse season (one solar eclipse) and then a three-eclipse season (two solar):

	New Moon	solar eclipse in series
2000	Feb. 5	150
	Mar. 6	.
	Apr. 4	.
	May 4	.
	June 2	.
	July 1	117
	July 31	155

As before, the series 6 lunations apart have the relation +5. But the series only 5 lunations apart have the relation −33, and the series only 1 lunation apart have the relation +38 (the sum of 12 and 26). All these relations hold throughout the system. So what will be the saros number of the next eclipse? It will be 6 lunations after that of July 1 (5 after that of July 31) and will be 117+5, or 155−33, that is, 122. And its date will be Christmas Day, 2000.

Remember that the two outer eclipses of a three-eclipse season must be slight ones (the middle one being very central, near to the actual node). In fact these flanking eclipses must be near the beginnings or ends of their series. Look again at the table for the year 2000. The series-number 117 is lower than most of the numbers occurring in our time. This is an eclipse near the end of an old series; it is in fact a partial eclipse in the Antarctic; the series is moving off the southern end of the Earth. Conversely the series number 155 is high: this is an eclipse near the beginning of its series, just moving onto the Arctic. (Both belong to the same season of solar eclipses at the ascending node—odd-numbered series—moving southward across the globe.)

The statement is true both ways around: all flanking eclipses in three-eclipse seasons are near the beginnings or ends of their series; all series begin and end with flanking eclipses in three-eclipse seasons.

What about numbering within a series? For example, the 1991 July 11 eclipse is number 36 of the 71 in solar series 136.

But, since the two ends of a series—such as eclipses 1 and 71 in this series—are insignificant or even doubtful, another method is numbering relative to the center. The 1991 July 11 eclipse is the center of its series, so we can give it the relative number 0. The eclipse after it becomes 1, those before it −1, −2 etc. back to the first eclipse, −35.

In 2000, the eclipse of July 1 is absolute number 68 of the 71 in series 117; that of July 31 is 5th of the 71 in series 155. But relatively, July 1 is number 32: it is 32 eclipses after the zero or central eclipse of its series. And July 31 has the relative number −30: it is 30 eclipses before the central eclipse of its series. These numbers better express the fact that these two eclipses, flanking members of a three-eclipse season, are slight ones near the end and beginning respectively of their series.

Besides the inex of 358 lunations, Van den Bergh gave the name tritos to a period of 135 lunations, that is, one inex minus one saros. (It works out to 3986.6 days, or 11 years less about a month.) This is when the next eclipse in the next-numbered series comes. Thus, the 1991 July 11 eclipse is the central eclipse of series 136. One tritos later comes the next eclipse in series 137 (on 2002 June 10). But it is one place before the center of that series; the center of its series comes another saros later (on 2020 June 21).

And the semester of 6 lunations, the usual interval between eclipses in successive eclipse seasons (slightly longer than our eclipse half-year which is between the centers of eclipse seasons), is 5 tritos minus 3 saros, or 5 inex minus 8 saros.

All these intervals—lunation, semester, saros, tritos, inex—again remind us of music. (In music, intervals are built of each other, or are differences between each other: the fifth is a fourth plus a whole tone, or a major third plus a minor third . . . And the sounds closest together—by a semitone—are most distantly related and most dissonant, whereas there are wide intervals between sounds most closely related—the octave, the twelfth.)

Van den Bergh and Meeus in their books have large "Panoramas" of eclipses: tables in which each column is a saros series. The eclipses in the next column are an inex or a tritos away (that is, in Van den Bergh's panorama all the horizontal relationships are inex and all the diagonal ones tritos, in Meeus's the other way around); and the semester is a kind of long knight's move away. These networks look like a kind of celestial harp.*

Example: solar saros series 136

Series 136 began in the year 1360. This does not mean that series are numbered by dividing their birth-years by ten, though it might say something about the series' luck. The timing of its first pass destined it to cross the Earth's almost exact center 631 years later—at its 36th eclipse.

Of its 71 eclipses, the first 8 (1360 to 1486) were partial ones; the next 6 (1504 to 1594) annular; the next 5 (1612 to 1685) annular-total; the main bulk of 45 eclipses (1703 to 2496) are total; and the last 7, from 2514 to 2622, will again be partial. What this means is that the series evolves in two main ways:

—Moving northward. At a pre-saros time (say, one saros before the first eclipse, in early June 1342) the Moon arrived at its descending node too early; it had dived too far south by the time it passed through the New Moon alignment. Arriving slightly later, in 1540 its shadow began to brush the Earth's southern end. But the axis of the shadow still missed, until in 1564 it began to intersect the Earth. Descending node and New Moon steadily squeezed closer together, until at eclipse number 36 of the symmetrical 71—in 1991—they coincide, yielding the most central solar eclipse in many centuries. From then on, the arrival at descending node us after the New Moon moment, so that the axis sweeps farther north, until from 2514 onward it misses.

---

* What I really means is that they remind me of my diagram of the imaginary musical instrument I call a *Plurry*, in my pamphlet of that title.

—Moving nearer. At the time when the series opened, the Moon was beyond its average distance. At each successive eclipse, the distance became shorter, until in our time the eclipses of this series are at very near New Moons. After that the Moon-distance expands again, so that the paths of totality become narrower.

The two factors—centrality and nearness—approach their climaxes—that is, their minima—in the twentieth and twenty-first centuries:

	gamma	dist.	duration
1901	-.362	56.38	6m 28s
1919	-.294	56.30	6m 50s
1937	-.224	56.24	7m 04s
1955	-.151	56.18	7m 07s
1973	-.076	56.13	7m 03s
1991	-.002	56.09	6m 53s
2009	.072	56.06	6m 39

"Gamma" is the measure of centrality. In 1901 the axis of the shadow passed south of the Earth's center by 0.362 of the Earth's radius of 6378 kilometers. "Distance" is that of the Moon's center from the Earth's center at mid eclipse, in Earth-radii. It does not reach its minimum (56.01) until the eclipse of 2063. "Duration" is the longest reached, at some point along the eclipse path. It is caused by the width of the umbra at the Earth's surface and its speed over the surface. So, though primarily determined by centrality and distance, it is affected also by the latitude and by the position of Earth and Moon in their orbits. The net result is that the six twentieth-century eclipses in this series are the century's longest, including the only ones with duration over 7 minutes until the year 2168.

In the maps of the series' paths, there are some obvious regularities. The first few paths, the annular ones, become narrower (and their narrowest parts are in their middles); narrowest are the annular-total paths. Then the total paths begin with that of 1703, and they grow broader. All this is because of the Moon's decreasing distance. After the 21st century the Moon becomes more distant and the paths narrower. Superimposed is the broadness caused by the glancing angle of the shadow at the polar eclipses, especially the extreme ones in 1504 and 2496.

As for the positioning of the paths, the two polar views have a swirling regularity of three directions (which would be clearer if we showed only the extreme 9 eclipses, instead of 12). And the central

Tracks of all the central eclipses of solar series 136 (annular eclipses plain, annular-total asterisked, total bold). The bottom and top pictures are centered on the south and north poles, and show the first 12 and last 12 central eclipses. The middle three pictures, centered on the equator at longitudes 120° W., 0°, and 120° E., show each third eclipse, beginning with the first (1504), third (1540), and second (1522) respectively.

mass of each three-set (from, say, 1775 to 2045, or 1757 to 2027, or 1739 to 2009) shows the most impressive regularity: the march of the great totality-paths northward up the eastern Pacific and North America, up Africa and Europe, and up Australasia and east Asia.

But the transitions between these zones look so confusing that one is tempted to reconsider the validity of the saros. From 2063 to 2171 the northward progress of the tracks slows, stops, and reverses. Then from 2225 to 2334 they slew westward. After that they continue northward, but with the East Asia set now going up Europe, the Afro-European set going up North America, and the American set transferred to East Asia.

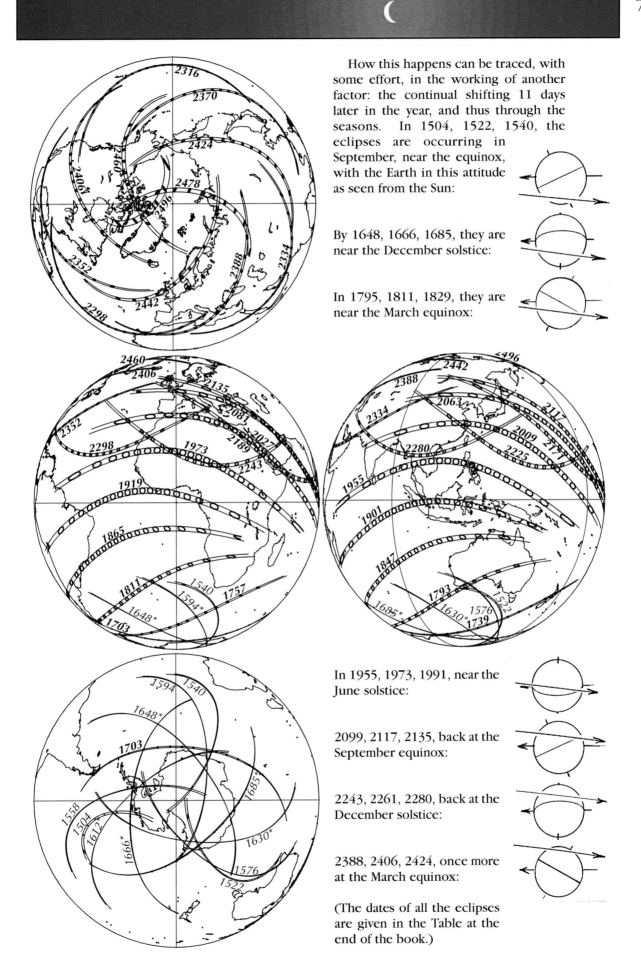

How this happens can be traced, with some effort, in the working of another factor: the continual shifting 11 days later in the year, and thus through the seasons. In 1504, 1522, 1540, the eclipses are occurring in September, near the equinox, with the Earth in this attitude as seen from the Sun:

By 1648, 1666, 1685, they are near the December solstice:

In 1795, 1811, 1829, they are near the March equinox:

In 1955, 1973, 1991, near the June solstice:

2099, 2117, 2135, back at the September equinox:

2243, 2261, 2280, back at the December solstice:

2388, 2406, 2424, once more at the March equinox:

(The dates of all the eclipses are given in the Table at the end of the book.)

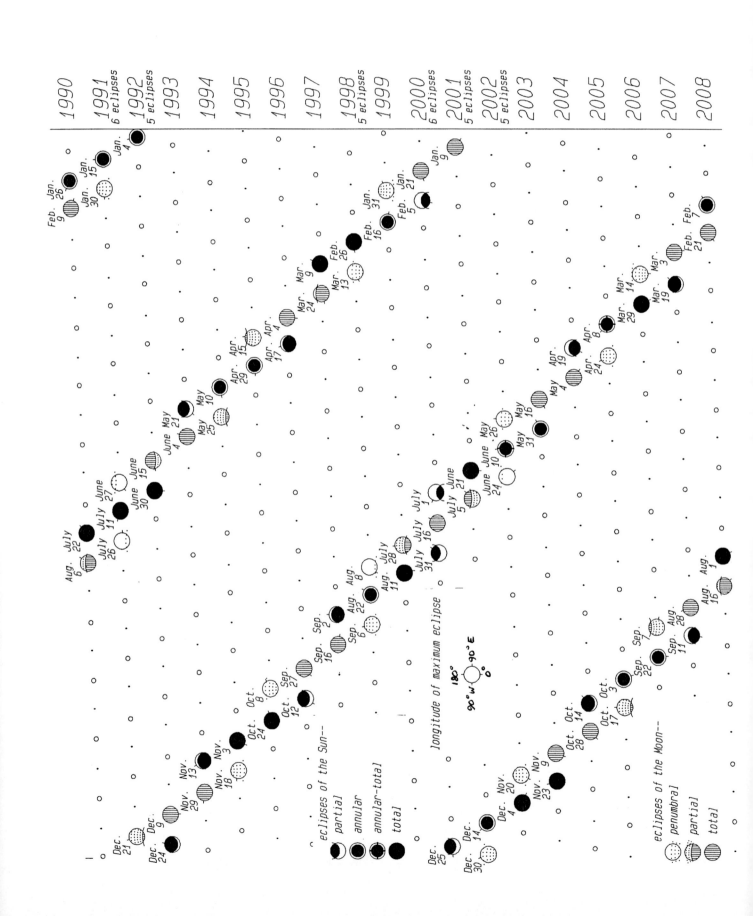

eclipses of the Sun--
partial
annular
annular-total
total

eclipses of the Moon--
penumbral
partial
total

longitude of maximum eclipse

180°
90° W
90° E
0°

1990
1991 6 eclipses
1992 5 eclipses
1993
1994
1995
1996
1997
1998 5 eclipses
1999
2000 6 eclipses
2001 5 eclipses
2002 5 eclipses
2003
2004
2005
2006
2007
2008

Pattern of Eclipses—the "Bead Curtain" chart

This time-graph, which begins overleaf and runs for twelve pages, should be read as a continuous stream by turning the book sideways.

An earlier version appeared in editions of my *Astronomical Companion*. In that version the symbols are pictures of the eclipse as seen by us on the Earth: that is, they show the Sun or Moon at maximum eclipse. Wanting to redo it for a span of two centuries, and not to have to fill in by hand the black overlaps for partial solar eclipses and the mists of dots for the Moon's penumbra, I spent a week extending the computer program with involved trigonometry. I can't resist including one page of the chart (opposite) in its abandoned format.

Only after then plotting the whole two centuries did I think of a radically different style, less simple and perhaps less pretty, but more informative; and had to spend another week reprogramming.

Along the line representing each year, time runs from right to left. This relates better to the eastward (leftward, as seen from the northern hemisphere) travel of solar-system bodies; each Full Moon appears farther east (left) against the starry background; the Earth circles the same way as seen from the Sun.

Each syzygy (New or Full Moon) is plotted with a small symbol; or, if an eclipse occurs at it, the symbol is enlarged and packed with quite a lot of information. It is a miniature picture looking outward from the Sun and showing not the eclipse as seen from Earth, but the thing traversed by the Moon: Earth, or Earth's shadow.

At each New Moon, the circle is the Earth; across it, if there is an eclipse, an arrow represents the path of the Moon's umbra, and shading represents the region traversed by the Moon's penumbra. The arrow is drawn thin if the umbra does not reach as far as the Earth, and thick if it reaches past the Earth's center. If the umbra reaches as far as the Earth's surface, but not as far as the center, the arrow has a short thick segment in the middle. This umbra arrow may miss the Earth, only the penumbra touching (a partial eclipse, denoted by "*pa*"). If it crosses some part of the Earth, the eclipse may be annular ("*an*") with thin arrow; total ("*TO*") with thick arrow; or annular-total ("*aT*"), with arrow thick only in the middle. (In this case the length of the thick part of the arrow is not accurate, only symbolic. Some annular-total eclipses are total for almost the whole path, some for only a few miles in the middle, but we show them all the same way.) The device of thin and thick arrows is used even for partial eclipses, so that one can see whether the eclipse *would* have been annular, annular-total, or total *if* it had been aimed more centrally at the Earth.

A year is written with bold type if it contains at least one chance to see the Sun totally eclipsed.

At each Full Moon, the circular patch represents Earth's shadow, with its two parts: inner dark umbra, outer faint penumbra. Usually the Moon passes above or below this shadow, and is not shown. But if there is an eclipse, then in enlarged view the Moon's path across the shadow is shown as an arrow, with the Moon itself as a small circle (to scale) at mid eclipse. The Moon may touch the penumbra only (penumbral eclipse, "*pe*") or also the umbra (partial eclipse, "*Pa*") or become immersed in the umbra (total eclipse, "*TO*"). Here too we use a thicker arrow to show when the Moon is nearer than average, though this makes less difference with lunar eclipses (it merely means that the Moon appears larger).

The offset of the arrow from the center of the symbol represents the quantity called γ (*gamma*): the distance by which the axis of the Moon's shadow misses the center of the Earth; or, in a lunar eclipse, the distance by which the Moon misses the center of the Earth's shadow. This is the main factor distinguishing central from peripheral eclipses, total from partial etc.

The horizontal level represents the plane of Earth's orbit,76 the ecliptic. Thus the arrows are straight, approximately as an observer at the Sun would see the path of the Moon and its umbra across the Earth. An eclipse happening at the ascending node makes a path sloping northward (upward), one at the descending node southward, both at the 5° angle of the Moon's orbit. **continued after the chart**

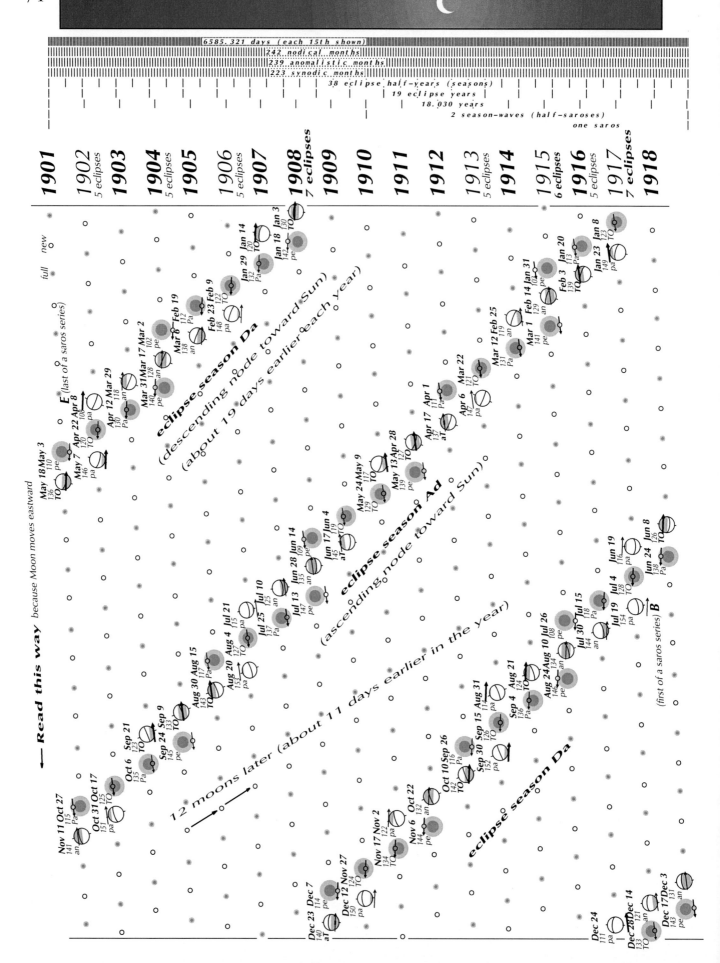

6585.321 days (each 15th shown)
242 nodical months
239 anomalistic months
223 synodic months
38 eclipse half-years (seasons)
19 eclipse years
18.030 years
2 season-waves (half-saroses)
one saros

1901 **1902** *5 eclipses* **1903** **1904** *5 eclipses* **1905** **1906** *5 eclipses* **1907** **1908** *7 eclipses* **1909** **1910** **1911** **1912** **1913** *5 eclipses* **1914** **1915** *6 eclipses* **1916** *5 eclipses* **1917** *7 eclipses* **1918**

full *new*

— **Read this way** *because Moon moves eastward*

eclipse season Da
(descending node toward Sun)
(about 19 days earlier each year)

eclipse season Ad
(ascending node toward Sun)

12 moons later (about 11 days earlier in the year)

eclipse season Da

E (last of a saros series)

B
(first of a saros series)

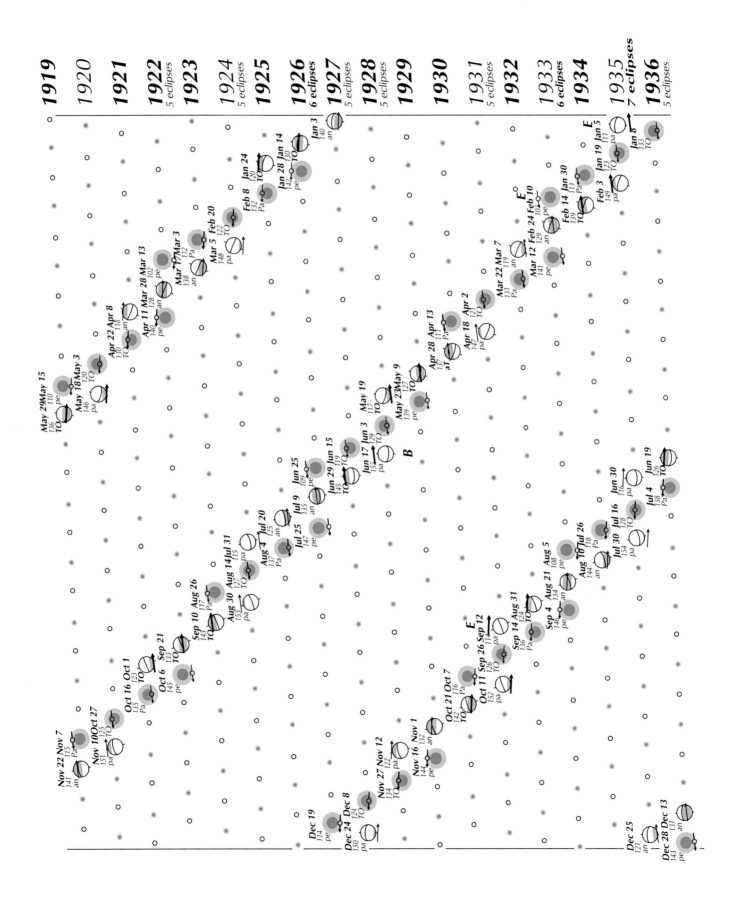

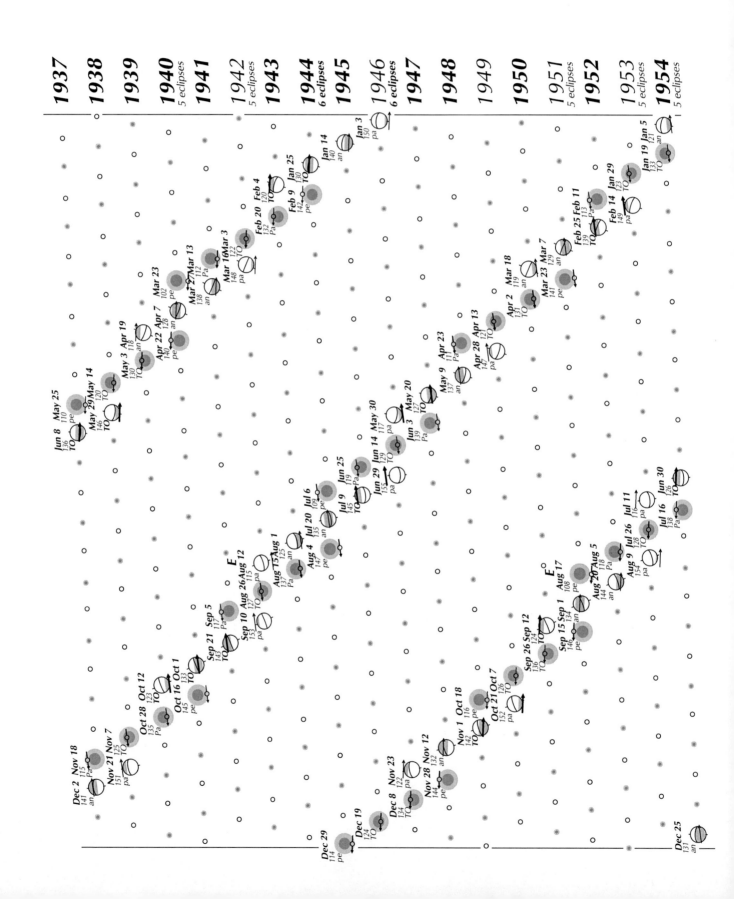

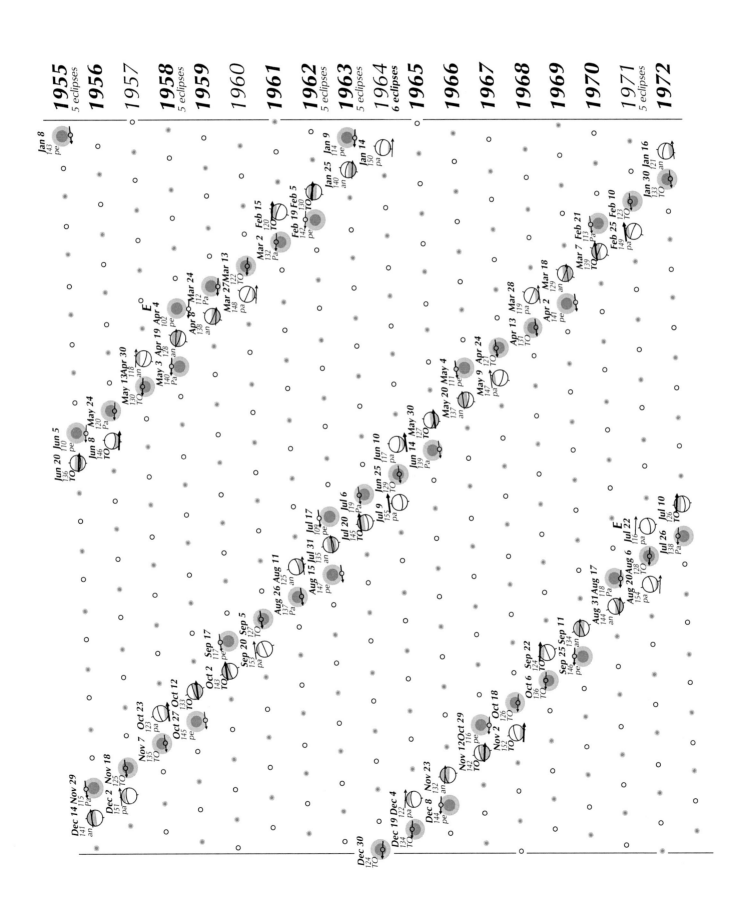

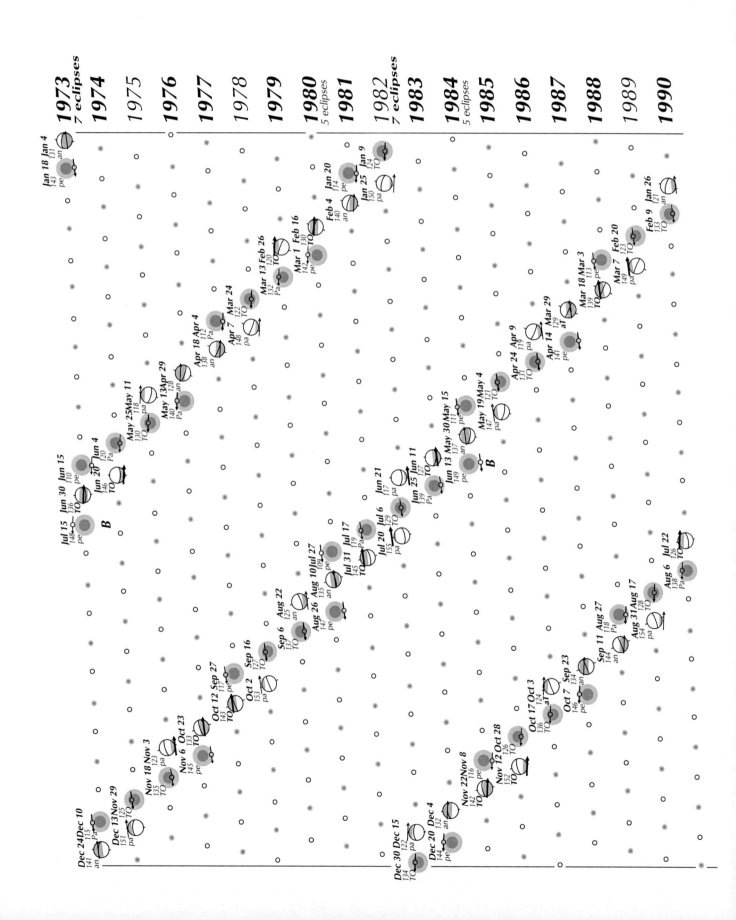

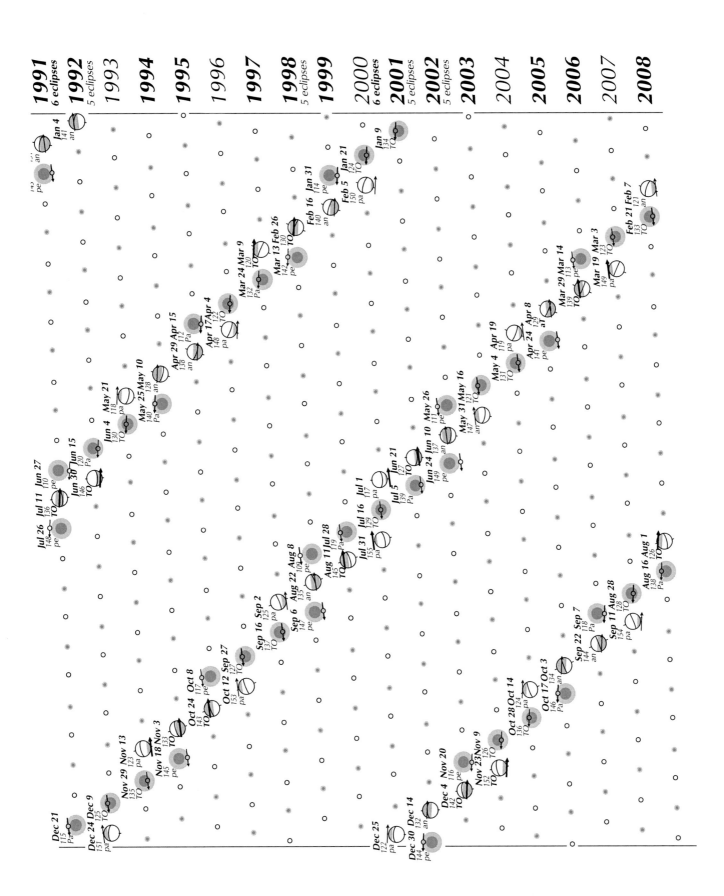

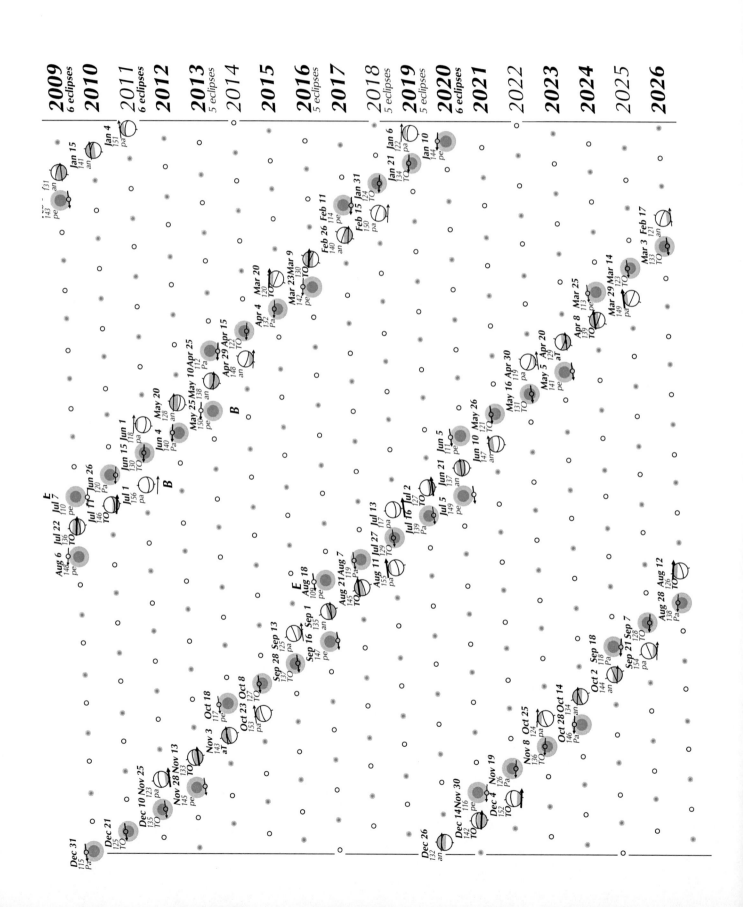

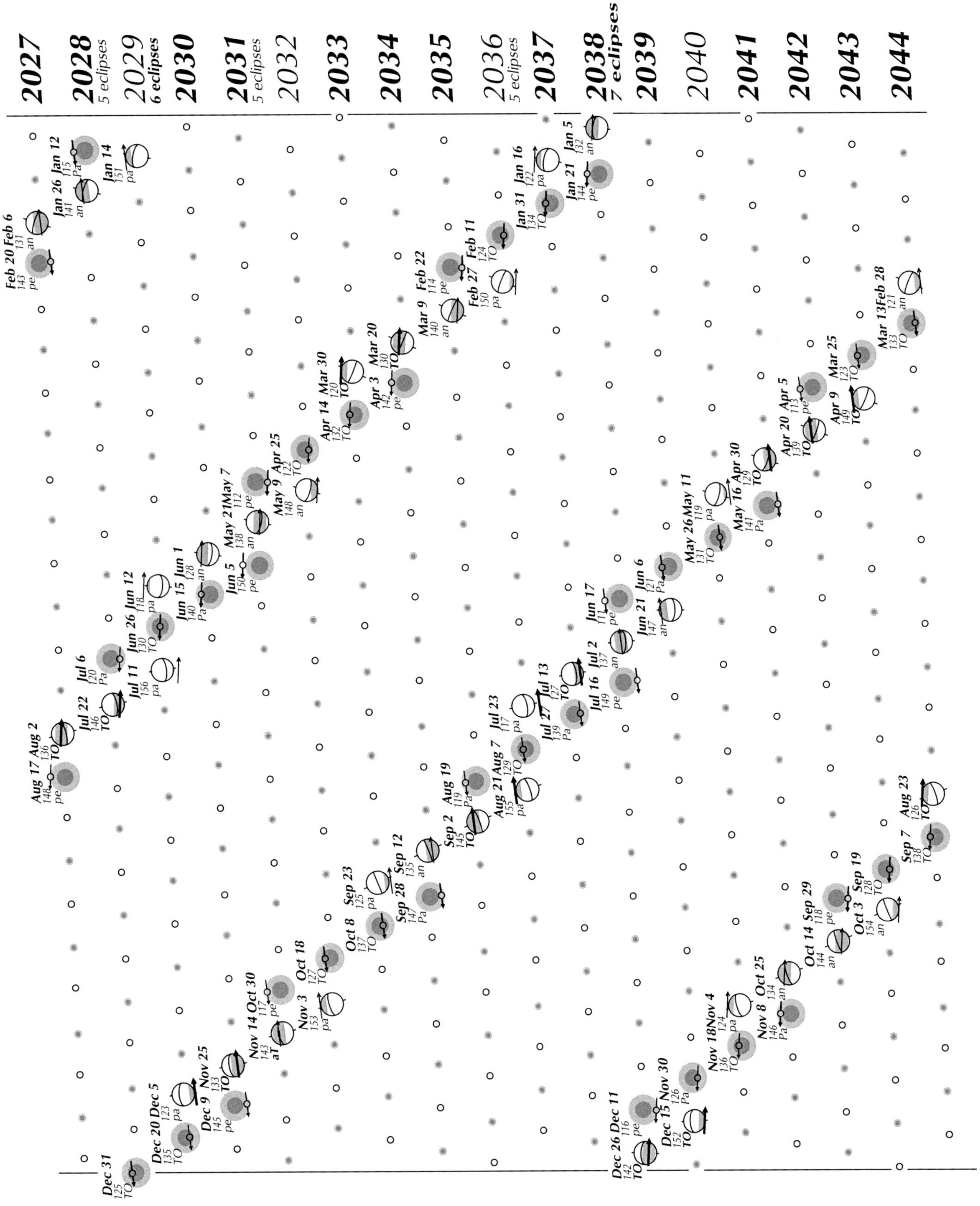
2027
2028
5 eclipses
2029
6 eclipses
2030
2031
5 eclipses
2032
2033
2034
2035
2036
5 eclipses
2037
2038
7 eclipses
2039
2040
2041
2042
2043
2044
Feb 20 143 pe
Feb 6 131 an
Jan 26 141 an
Jan 12 115 Pa
Jan 14 151 pa
Aug 17 148 pe
Aug 2 136 TO
Jul 22 146 TO
Jul 11 156 pa
Jul 6 120 Pa
Jun 26 130 TO
Jun 12 118 pa
Jun 15 140 Pa
Jun 1 128 an
Jun 5 150 pe
May 21 138 an
May 7 112 pe
May 9 148 an
Apr 25 122 TO
Apr 14 132 TO
Mar 30 120 TO
Apr 3 142 pe
Mar 20 130 TO
Mar 9 140 an
Feb 22 114 pe
Feb 27 150 pa
Feb 11 124 TO
Jan 31 134 TO
Jan 16 122 pa
Jan 21 144 pe
Jan 5 132 an
Dec 31 125 TO
Dec 20 135 TO
Dec 5 123 pa
Dec 9 145 pe
Nov 25 133 TO
Nov 14 143 aT
Oct 30 117 pe
Nov 3 153 pa
Oct 18 127 TO
Oct 8 137 TO
Sep 23 125 pa
Sep 28 147 Pa
Sep 12 135 an
Sep 2 145 TO
Aug 19 119 Pa
Aug 21 155 pa
Aug 7 129 TO
Jul 23 117 pa
Jul 27 139 Pa
Jul 13 127 TO
Jul 16 149 pe
Jul 2 137 an
Jun 17 111 pe
Jun 21 147 an
Jun 6 121 Pa
May 26 131 TO
May 11 119 pa
May 16 141 Pa
Apr 30 129 TO
Apr 20 139 TO
Apr 5 113 pe
Apr 9 149 TO
Mar 25 123 TO
Mar 13 133 TO
Feb 28 121 an
Dec 26 142 TO
Dec 11 116 pe
Dec 15 152 TO
Nov 30 126 Pa
Nov 18 136 TO
Nov 4 124 pa
Nov 8 146 Pa
Oct 25 134 an
Oct 14 144 an
Sep 29 118 pe
Oct 3 154 an
Sep 19 128 TO
Sep 7 138 TO
Aug 23 126 TO

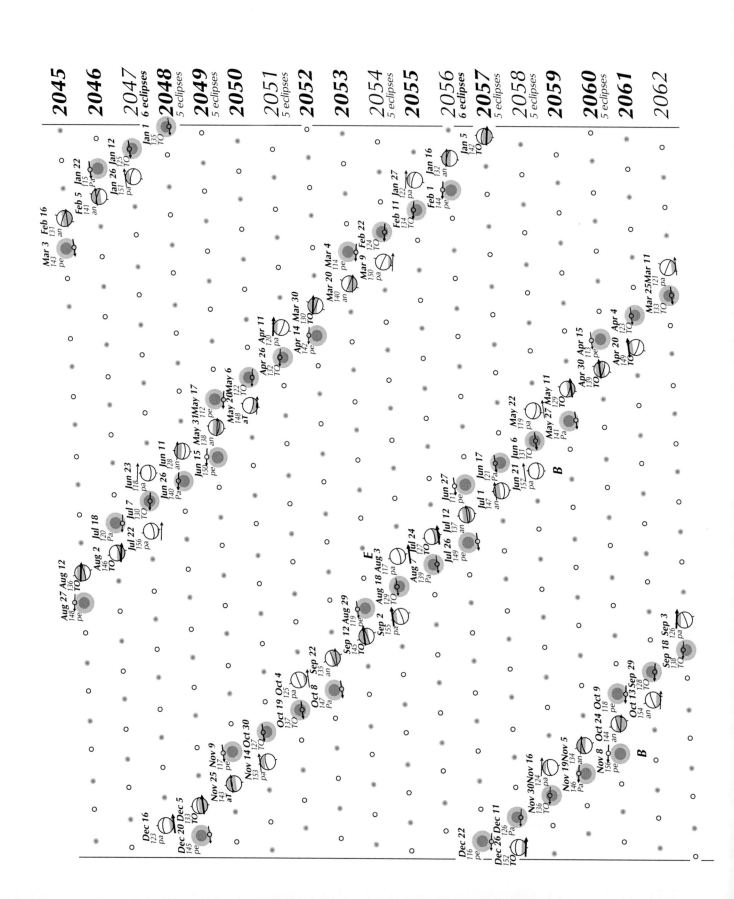

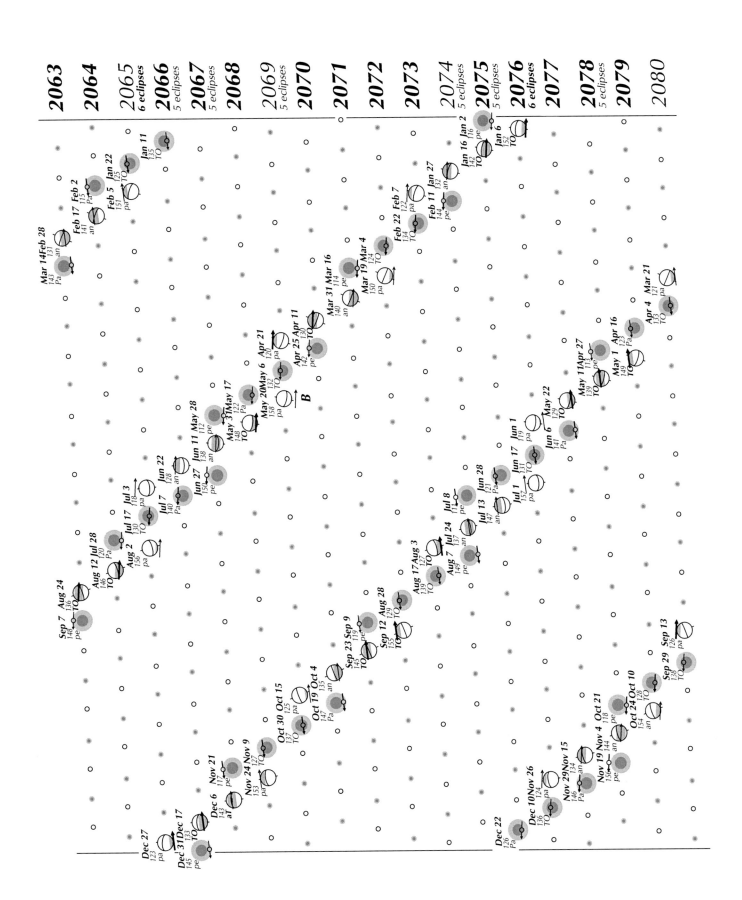

2063 **2064** **2065** 6 eclipses **2066** 5 eclipses **2067** 5 eclipses **2068** **2069** 5 eclipses **2070** **2071** **2072** **2073** **2074** 5 eclipses **2075** 5 eclipses **2076** 6 eclipses **2077** **2078** 5 eclipses **2079** **2080**

Jan 11
135
TO

Jan 22
125
TO

Feb 2
115
Pa

Feb 5
151
pa

Jan 2
116
pe

Jan 16
142
TO

Jan 6
152
TO

Feb 17
141
an

Feb 28
131
an

Jan 27
132
pe

Feb 11
144
TO

Feb 7
122
pe

Feb 22
134
TO

Mar 14
143
Pa

Mar 4
124
pe

Mar 19
150
pa

Mar 21
121

Mar 31
140
an

Mar 16
114
pe

Apr 4
133
TO

Apr 11
130
TO

Apr 25
142
pe

Apr 16
123
Pa

Apr 21
120
pa

May 1
149
TO

May 6
132
TO

May 11
139
TO

Apr 27
113
pe

May 20
158
pa

B

May 31
122
pe

May 17
112
an

May 22
129
TO

Jun 6
141
Pa

Jun 11
148
an

Jun 1
119
pe

Jun 27
154
pe

Jun 22
128
an

Jun 17
131
TO

Jul 7
140
Pa

Jun 28
121
an

Jul 17
130
TO

Jul 3
120
pa

Jul 1
152
TO

Jul 13
147
an

Jul 8
111
pe

Aug 2
156
pa

Jul 28
146
TO

Aug 7
145
pe

Jul 24
137

Aug 12
TO

Aug 17
139
an

Aug 3
122

Aug 24
136
TO

Sep 7
148
pe

Sep 12
155
TO

Aug 28
129
TO

Sep 23
145
TO

Sep 9
119
pe

Sep 29
128
TO

Sep 13
126
pa

Oct 19
147
Pa

Oct 4
135
Pa

Oct 30
137
TO

Oct 15
134
pa

Oct 24
154
an

Oct 10
128
an

Oct 21
118
pe

Nov 9
127
TO

Nov 24
153
pa

Nov 21
117
pe

Nov 19
156
pe

Nov 4
144
an

Dec 6
143
aT

Nov 29
146
pe

Nov 15
134

Nov 26
124
pa

Dec 10
136
TO

Dec 27
123
pa

Dec 17
133
TO

Dec 31
145
pe

Dec 22
126
Pa

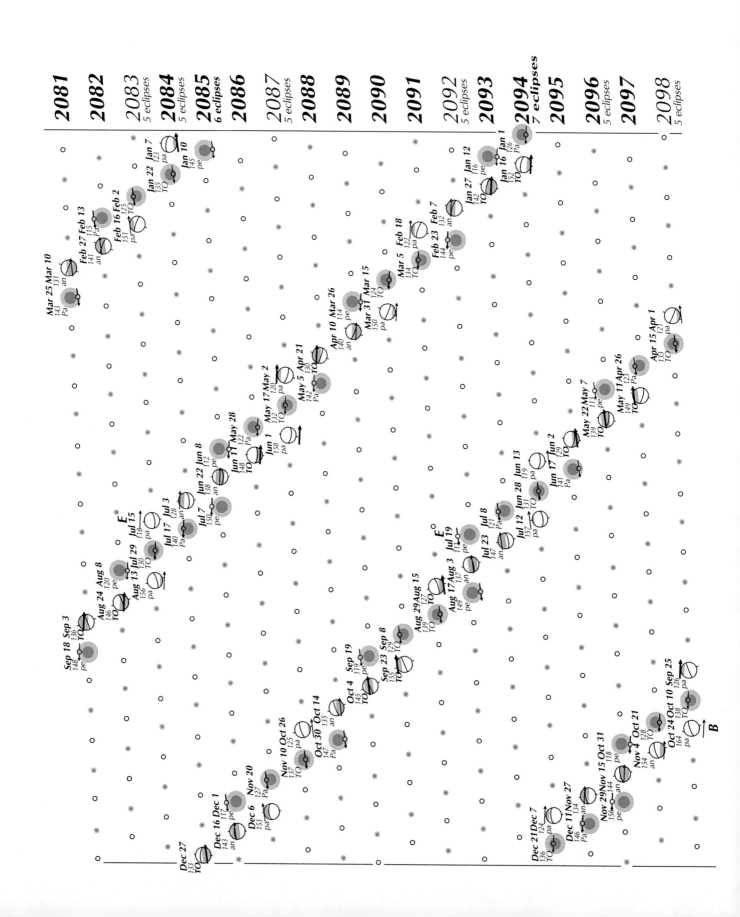

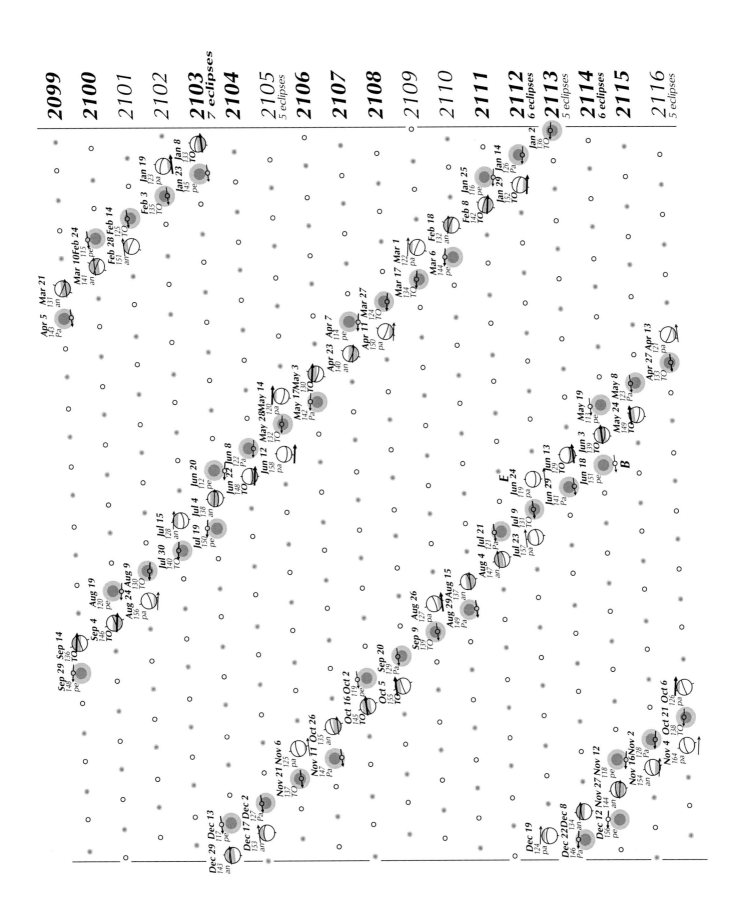

continued from before the chart

Each Earth symbol is marked with its north and south poles and its equator. Thus you can see how the Earth looks from the Sun's direction at each eclipse. Around March 21 the Earth is leaning north-pole backward as it progresses (leftward) in its orbit; around June 21 the north hemisphere is tipped toward the Sun; around September 23 the north hemisphere leans forward; and around Dec. 22 the north hemisphere leans away. The continuous variation from one attitude to the next can be followed through groups of eclipses. This is about as detailed as we can get on this scale. The Earth is rotating rightward (the same way as the arrows of the eclipse paths though more slowly), and if it were possible to show longitude-lines or a map it would be seen in any stage of rotation (any longitude facing the Sun) at the time of mid eclipse. The arrows of the eclipse paths, though virtually straight and at a steady angle in relation to the ecliptic, make paths with an unlimited variety of compound curvatures across the geography of the rotating and variously tilted Earth.

The label by each eclipse includes, besides its date and type, the number of the saros series to which it belongs. A *B* below an eclipse means that it is the first of its series (as for 1917 July 19); an *E* above it means that it is the last of its series (as for 1902 Apr. 8).

There are 18 years to a page, matching as closely as possible the 18.03-year length of the saros. And thus *each page approximately repeats the preceding one. For every eclipse (except the "B" and "E" ones) there is a similar eclipse in nearly the same position on the next and previous pages!*

Follow a series down the pages, such as series 136 starting at the top middle of the first page, and watch how it gradually changes and gradually drifts in time.

The nearest we can get to a century on this system is 6 pages, or 108 years. The unit of 6 saroses, or 108.18 years, might be called a "saros century." We show about two eclipse centuries, from 1901 to 2116. (Centuries, in the regular sense, start with the years 1, 101, 201 . . . 1901, 2001, and so on.)

The "comb" diagram over the first page shows the many units of time whose near-coincidence forms the saros: the synodic, draconic, and anomalistic months, as well as the non-lunar units of the day and year. A unit not mentioned before, but which the chart suggests, is what we might call the halfwave or seasonwave, or simply half saros, or perhaps the saros decade (of which there are 12 in the saros century): the rhythm in which the bands of the eclipse seasons come to the side of the chart (the beginnings of the years). In other words, it is the interval between years in which eclipses are occurring at the same time of year. In 1908, eclipses occur in January, so 9 years later in 1917 there are again eclipses in January—though at the opposite nodes. Only after a complete saros, in 1926, do *similar* eclipses occur in January.

Using more vertical space, we could make the chart show why eclipses do not happen at the other syzygies (New and Full Moons) between eclipse seasons:

 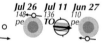 

Here the arrows show the Moon passing up to 10 Earth-widths north or south of the Earth or its shadow. Notice how the tilt of the arrows gradually changes between eclipse season Ad (in January and December), when the Moon's orbital plane is "downward to the left" (southward to the west), and eclipse season Da (in July).

One way of treating the chart is as a visual index to the many solar-eclipse globes and lunar-eclipse diagrams that appear elsewhere (such as in the *Astronomical Calendar*).

Another is to sit back and contemplate it and to swing the fist in a horizontal circle (counterclockwise as seen from above), and think of the Moon in its continually re-tilting orbit, alternately shadow-swiping and being shadow-swiped by the Earth.

But the more you gaze at this Bead Curtain, the more it shimmers—the more cross-patterns are discerned in the pattern.

The first extensive feature that stands out is that of the bands formed by the eclipse seasons, running down the chart but shifting backward through the years. Though there seem to be many bands there are really only two, perpetually slanting from December to January and reappearing in December, so as to spiral down the whole length of the chart: the seasons we have called "Ad," in which the ascending node of the Moon's orbit is toward the Sun, so that solar eclipses happen at it and lunar eclipses at the descending node; and the opposite "Da" seasons nearly half a year later (descending node toward the Sun, solar eclipses at it and lunar eclipses at the ascending node).

A line along the center of each band would connect the instants in the years when the line of nodes of the Moon's orbit is oriented exactly along the Earth-Sun line. Eclipses are more central as they fall closer to this axis or equator of the band, more marginal as they lie at the band's fringes—marginality in these bands is also marginality as an eclipse. (Flanking lines could also be drawn, beyond which eclipses cannot occur.)

In any single eclipse season, an eclipse near its center is a more or less central one and those at the edges are near opposite extremes of marginality. However, the pattern is symmetrical only if one eclipse is *very* central. For example 1991 June 27, July 11, July 26: penumbral lunar eclipses at the south and north edges of the shadow, flanking the almost perfectly central solar eclipse. It coincides (to 3 minutes) with the node passage, therefore with the exact center of the eclipse season, so the other two eclipses are as far as possible before and after the passage of the opposite node. Likewise the very central 3-eclipse season in 2000, this time with a lunar eclipse in the middle. But other 3-eclipse seasons, such as that in 1998, are not at all symmetrical: that is to say, the middle eclipse being somewhat un-central, the flanking ones have the opposite asymmetry to what you might expect. And if the season consists of only two eclipses (straddling the center of the season), then they are both either southerly or northerly. For example 1991 Jan. 15 and Jan. 30: a solar eclipse across the southern hemisphere of the Earth, followed by a lunar eclipse across the southern part of the Earth's shadow. The principle is not symmetry, but the timing of the Moon's motion past the node. An eclipse before the middle of a season takes place when the Moon is approaching one node; an eclipse after the middle of a season takes place when the Moon has just passed the opposite node—and is therefore now on the same side of the ecliptic. Sometimes this rule barely holds: one of the eclipses is only just in the right hemisphere, as on 1998 Feb. 26. Such a 2-eclipse season comes close to being a 3-eclipse one: the Moon's shadow barely misses the Earth on Feb. 11, so Feb. 26 is not long before the center of the season. The 1998 3-eclipse season comes close to being a 2-eclipse one: Aug. 22 and Sep. 6 are on either side of the center, Aug. 8 almost misses being an eclipse.

The most marginal eclipses of all are those which are at the beginnings or ends of their saros series. An initial eclipse (marked by a *B* below) is always at the end of a 3-eclipse season; a terminal eclipse (marked by an *E* above) is always at the beginning of a 3-eclipse season—fortunately for the placing of these symbols! Old eclipses, that is, eclipses toward the end of old series, line the upper edges of the bands, young eclipses form the lower fringes; each season progresses from old through mature (series-central) to young eclipse.

Some 2-eclipse seasons have the order solar-lunar, some lunar-solar; a 3-eclipse season always effects a changeover from one to the other (though the changeover can also happen without a 3-eclipse season, as between 1992 and 1993).

From season to season, eclipses tend to be of complementary types. Thus in 1994, May 10 is annular across the northern hemisphere, Nov. 3 total across the southern hemisphere; May 25 partial through the northern shadow, Nov. 18 penumbral through the southern shadow.

Then there are the regularities that run down the eclipse season bands. The bands are like long ropes made of short fibers: they are composed from many overlapping rows of lunar and of solar eclipses. Each row consists of eclipses that are 12 lunations apart (the nearest they can come to being a solar year apart). These rows, like the bands they compose, are constituents of the pattern of eclipses perhaps as important as saros series and eclipse seasons are, for they seem as necessary in describing or making regularity out of the welter of eclipses.

The rows consist typically of 4 eclipses, often 5, rarely 3. Slight eclipses, those that come nearest to not being eclipses, are first or last in their rows. If a row is 5 long, its end eclipses are more marginal, its central one more central. 3-eclipse seasons often (not always) begin or end 5-eclipse rows. Terminal eclipses of saros series are always at the top of 5-eclipse rows, and initial eclipses are at the bottom of such rows (though by no means all eclipses at the limits of 5-eclipse rows are saros-initial or terminal).

Down each row there is a progression from near one extreme to another. Consider the lunar row 1998 Aug. 8, 1999 July 28, 2000 July 16, 2001 July 5, 2002 June 24: penumbral at the north edge of the shadow, partial at the north, total, partial at the south, penumbral at the south. The first is so slight that it is the terminal eclipse of its series; the middle eclipse of the five is very central. And consider the solar row 2000 July 1, 2001 June 21, 2002 June 10, 2003 May 31: partial at the south pole, total across the southern hemisphere, annular across the northern hemisphere, annular near the north pole. Any row of eclipses (lunar or solar) at the descending node passes successively farther south, at the ascending node farther north. (This is in contrast with successive eclipses in a saros series, which if at the descending node migrate successively northward, and vice versa.) When there are only 4 in a row, the end ones are generally not so marginal nor the middle ones so central: total eclipses (though comparatively slight, Arctic or Antarctic ones) may stand at the ends of rows, such as on 1985 Nov. 12.

Along each row there is also a progression in the Moon's distance. Thus 2000 July 1, 2001 June 21, 2002 June 10, 2003 May 31: partial (but near enough to be total), total, annular (very close to being annular-total), and annular. In one of the lunar rows flanking this (2002 May 26 etc.) the progression is similar, the Moon gradually becoming more distant; the other flanking row (1998 Aug. 8 etc.) is so long that the progression reverses, the Moon at first becoming more distant and then closer; in the row 1997 Sep. 2 etc. the progression is from farther to nearer.

In the middle of the two-century chart is a row of only three: 1990 July 22, 1991 July 11, 1992 June 30. Such a short row that all three are total, and the middle one is the most central of all solar eclipses in the two centuries of our chart.

A further thought here is that this 3-eclipse row must verge on being a 5-eclipse row. That is, there must be near-miss eclipses just before and after it, at the New Moons of 1989 Aug. 1 and 1993 June 20. And if that is so, those near-eclipses must be just off the ends of series; that is, we will find a terminal eclipse in the saros position 18 years before the first of them, and an initial eclipse in the saros position 18 years after the other. And we do: 1971 July 22, and 2011 July 1.

Here we realize that not just single eclipses but seasons, rows, and even more extensive groups of eclipses participate in the saros rhythm. They have their near-repetitions 18 years earlier and later. The 3-row centered on the eclipse of 1991 July 11 has the saros numbers 126, 136, 146. It is reflected eighteen years earlier in a 4-row with the numbers 116, 126, 136, 146, and eighteen years later in a 4-row with the numbers 126, 136, 146, 156. The 1991 eclipse (the summit of saros series 136) is situated exactly in the middle of the 40-year gap between the death of series 116 and the birth of series 156. These are yet more of the symmetries of which this great eclipse—because of its own near-perfect centrality—is a center.

What about the vertical relation between the eclipse season bands? Between alternate bands it is of course the relation of the saros: the whole "Da" band sloping from December 1982 to January 2001 is repeated (as well as continued) in the next "Da" band sloping from December 2000 to January 2020. But is there a relation with the "Ad" band in between? In other words, is there a relation between eclipses a halfwave apart, other than that they fall at the same season of the year? Indeed, *are* there eclipses exactly a half-saros after other eclipses, that is, halfway between two eclipses of a saros series? Yes, there are, and they are similar, *except in the most important respect*. For, the saros being an odd number of lunations, 223, its half is a fraction, 111.5; therefore the eclipse a half-saros from a solar eclipse is a lunar one. Halfway between the nights of 1982 July 6 and 2000 July 16, when total lunar eclipses of series 129 occur, falls the eclipse in the day of 1991 July 11: total, and likewise very central, but solar. As for the parts of the Earth over which the eclipses take place, in longitude they roughly inter-leave each other; in latitude the lunar eclipses are south of the equator where the solar ones are north, because in July when the Sun is north of the equator the Full Moon is south of it. Look at the whole middle season of 1991 and that of 2000: the arrows slope the opposite ways, being at the opposite nodes, but they are at the same heights; but where in 1991 they cross the Earth in 2000 they cross its shadow. In other words the Moon, having passed at a certain level around one side of the Earth, passes 9.015 years later at a similar level around the other side. You can trace this relationship on each page of the chart.

This relation between eclipses only 9 years apart might have been easier to discover than the saros-relation between eclipses 18 years apart. The eclipse of the Moon in 1982 was easy to see over a wide area which included most of the path where the total solar eclipse of 1991 was seen. Having once noticed such a 9-year-5-day step from lunar eclipse to solar, or from solar to lunar, one could put them together to predict an 18-year-10-day step from solar to solar or lunar to lunar.

In such ways and more, the whole pattern can be dissected into regularities along different dimensions, all of which are dimensions of time. Yet all these regularities must tie tightly together, and are expressions of the underlying geometry of the Moon's motions. What at first seems a jumble of Beads dangling in the wind and offering contrasting faces from a mere caprice to be decorative turns out to be so solidly rule-bound that in the end the pattern itself is the only adequate expression of its own rules.

Even the saros-series numbers which dryly pepper the chart—and which are its only artificial feature—are absorbing in their regularities. We gave the rules for them before, but in the chart they are easier to survey. Across each eclipse season the numbers only *increase* (never decrease) because eclipses at the beginnings of seasons are toward the ends of old series, therefore in lower-numbered series; season-ending eclipses are near the beginning of younger, therefore higher-numbered series. The numbers increase by 12 from solar to lunar eclipse, and by 26 from lunar to solar; thus by 38 from solar to solar or lunar to lunar (in any three-eclipse season, where eclipses of the same kind occur only one lunation apart). From one season to the next, the number between corresponding eclipses increases by 5. (These eclipses are 6 lunations—a "semester"—apart.) Down the slanting rows within the eclipse seasons, the numbers increase by 10. These eclipses are 12 lunations apart, a rough lunar year. When an eclipse season reaches the junction of the years, the eclipses 10 saros numbers and 12 lunations apart get into the same calendar year and thus the same row of our chart: 1908 Jan. 3 and Dec. 23, 1917 Jan. 8 and Dec. 28 . . .

It is fairly easy to locate on any page the eclipse in any particular series, or to determine that the series has already ended or has not yet begun. Why? Because on every page can be found all twenty kinds of row: the lunar 0-row and the solar 0-row, the lunar 1-row and the solar 1-row, the lunar 2-row and the solar 2-row . . . So, if you find the 5-row, and see that it consists of eclipses in series 108, 118, 128, 138, and 148, you know that series 98 and older are dead, and series 158 and younger are not yet born.

There are many micro-patterns to be found. For instance, look at the symmetry in the year 1915 and its saros-repetition in 1933.

The Metonic Cycle

Here is something else traceable on the pattern-of-eclipses chart. In 432 B.C. Meton of Athens discovered by his own observations that Moon phases repeat on the same day of the year after 19 years.

```
years:             19 × 365.2425  = 6939.6075 days
synodic months:   235 × 29.53059 = 6939.6887 days
```

The difference is only 2 hours. Here are the dates of New Moon in two years:

	Jan.	Feb.	Mar.	Apr.	May	June	July	Aug.	Sep.	Oct.	Nov.	Dec.
1991	15	14	16	14	14	12	11	10	8	7	6	6
2010	15	14	15	14	14	12	11	10	8	7	6	5

Where the date is a day off, the cause may be the 2-hour imperfection, the variable lengths of synodic months, our definition of days as beginning at Greenwich midnight, or (most often) the leap-days in our present calendar. A consequence is that Februaries with only 3 Moon-phases recur at 19-year intervals (no Full Moon, for instance, in 1961, 1999, 2018, 2037) and so do months with a phase repeated (for instance a second Full Moon, now commonly called a "Blue Moon," as in March 1980, 1999, and 2018).

Meton's concern was that of a calendar-maker, trying to make lunar months fit into solar years. He proposed a calendar consisting of a 19-year cycle or "great year" into which were fitted 235 lunar months (110 of them with 29 days, 125 with 30); years 3, 5, 8, 11, 13, 16, and 19 had to have 13 months. He wanted his calendar to start at midsummer of 432 B.C., and it is sometimes said to have been adopted throughout Greece, but it may have been in actual use only at Athens from 338 to 290. Meton's 19 years are only 2 hours different from 235 lunar months. Callippus (about 350 B.C.) reduced the difference by combining 4 of Meton's 19-year cycles, slightly changed, into a 76-year-one; finally Hipparchus (about 130 B.C.), greatest of Greek astronomers, combined 4 of these into a 304-year cycle. The 19-year cycle, with Meton's name attached to it, lives on in the tables used for determining Easter. Meton also had a scheme of geometrical town planning; he was one of the intellectuals satirized in Aristophanes's comedy *The Clouds*."

Meton's 19-year cycle is not so strongly related to eclipses as the 18.03-year saros. But it is exactly a saros (223 lunations) plus the lunar approximation to a solar year (12 lunations). This gives ir a sliding relation to the pattern of eclipses. For example the New Moons on April 8 or 7 in 1902, 1921, 1940, 1959, and 1978 happen to bring solar eclipses. And they form a progression-but one with more rapid change than a true saros series, because they are in *neighboring* series. The series numbers increase by 10 (108, 118, 128, 138, 148) because this is also the rule for eclipses in a row, only 12 lunations apart. These eclipses are lined up vertically, but against the general saros slant they have to stay in place, as it were, by making an elongated knight's-move—hopping aside to the next series and then taking the saros leap along it. (The verticality of the Metonic series could be said to be what causes the slant of the true saros series.) Because each eclipse is a saros plus a year onward, it lands one line lower per page of our chart. This causes the Metonic series to drift down across the band of the eclipse season-the opposite motion to that of a true series, which begins with eclipses on the lower side of the band (end of the season). And thus this pseudo-series *begins* with an eclipse that is at the *end* of its true series, and moves back to eclipses in progressively older true series. Slipping rapidly across the band, the Metonic series soon finds itself out of it: at the next New Moon in the sequence, on 1997 April 7, there is no eclipse. Or, put another way, because the Metonic series changes more rapidly than a saros series it has a short lifetime: the path of the shadow progresses from the north pole to the south in 5 eclipses instead of 80. There continue to be New Moons on April 7 or 8 each 19th year, but not until many centuries have passed will they again lie within the eclipse season. (Meanwhile, interleaved with this sequence of New Moons, there is a similar Metonic sequence of Full Moons on the same date 19 years apart, but in different years; it begins to yield lunar eclipses on 2107 April 7.)

The dark lunar eclipses of Dec. 30 in 1963 and 1982, following the eruptions of Gunung Agung in Bali and of El Chichón in Mexico, were exactly a 19-year Metonic cycle apart.

Eclipse stories

B.C.:

2137 Oct. 22, solar annular, or **2133 Aug. 9, solar total?** The legendary Chinese emperor Yao sent two sages, Hsi and Ho, to the four corners of the Earth to keep the Sun on its course and prevent eclipses. Later, these Ministers of Astronomy neglected their duties and were drunk in their cities; they knew nothing on an occasion when "the Sun and Moon did not meet harmoniously in Fang." This was in the 5th year of emperor Zhong Kang (dated uncertainly around 2000), who sent an army to punish them, as described in a text called *The Punitive Expedition of Yin* (its leader). The way this story is sometimes told, Hsi and Ho (or "Hi and Ho") were astronomers hung high for being too drunk to predict an eclipse. It has been identified with one or other of these dates. Well, maybe.

Ian Dicks

1361 Feb. 15, lunar, : probably the earliest sure Chinese eclipse record.

1223 Mar. 5, solar total across Libya and Asia Minor. Ugarit was a city on the northern coast of Syria, lost until its site was discovered in 1928. One of the excavated clay tablets recorded an eclipse, with the month and the visibility of Mars. The eclipse was estimated to be that of 1375 May 3, but later research re-dated it to 1223.

763 June 15, solar total, across the northern end of Mesopotamia; deep partial in Assyria. According to a tablet: "Bur-Sagale of Guzana, revolt in the city of Assur. In the month Simanu, *shamash akallu*—the Sun was obscured." In 1867 Henry Rawlinson, "Father of Assyriology," identified this with the eclipse of 763; this is generally accepted and was of great importance in fixing Assyrian chronology. The prophet Amos, who lived around 750, referred, maybe, to an eclipse of the time: "I will cause the Sun to go down at noon, and I will darken the Earth in the clear day."

585 May 28, solar total. Cyaxares, builder of the newly powerful Median empire in Iran, attacked Alyattes II, of the Lydian kingdom in western Asia Minor. The armies met near the river Halys (Turkish: Kizil Irmak), but "Day was suddenly turned into night," and the kingdoms made peace. The account comes from the *Histories* (Book I, section 16) of Herodotus, written in Greek about 440. He added that the philosopher Thales of Miletus (about 624-546) had foretold the event, "fixing for it the very year in which it actually happened." The eclipse is presumed to be this one, whose central path ended in Asia Minor, so that the Sun would have become totally covered only just before setting. In 546, Cyaxares's great-grandson Cyrus the Great of Persia overthrew Alyattes's son Croesus.

480 Apr. 9, solar? Xerxes I, grandson of Cyrus and "king of kings" of the Persian empire, was on his way to his ill-fated invasion of Greece. He had had a bridge constructed across the Hellespont, from Asia to Europe. "At the beginning of spring," when his army had set out northward toward the bridge from Sardis, capital of the Lydian province in what is now western Turkey, "the Sun was unseen, leaving its seat in the sky, even though that was cloudless and very clear, and instead of day night came to be" (Herodotus, VII, 38). Xerxes had ignored less obvious omens of disaster (a cow giving birth to a hare); the Magi now told him that the Sun represented Greece and the Moon Persia and so he would conquer. A loyal and elderly follower, Pythius the Lydian, was more worried and begged to be allowed to leave the eldest of his five sons at home. Angered, Xerxes had the son cut in half and the army march between the pieces. Fortunately, the story may be untrue, since the total eclipse of 480 April 9 was visible across the southern Pacific, though that of Oct. 2 crossed northern Africa.

413 Mar. 4, lunar partial. On the night before the Athenian army, commanded by the rich, pious and ailing Nicias, was about to give up its disastrous siege of far-off Syracuse in Sicily and sail home, the Moon was eclipsed. Nicias insisted on delaying thrice nine days (a cycle of the Moon). The whole army was captured and left to starve in the quarries.

357 Feb. 29, solar total across the eastern end of the Mediterranean. Plutarch, in his biography of Dion, tyrant of Syracuse in Sicily, mentioned that a scholar named Helicon of Cyzicus predicted a solar eclipse while he was accompanying Plato on his third visit to Syracuse; and was rewarded by Dionysius with a talent—a potful of gold. The "tyrants" of Greek city-states were men who had managed to seize sole power; Dionysius the Elder of Syracuse was an example of the more oppressive kind, though with pretensions to the arts, as were his son Dionysius the Younger and his brother-in-law Dion, who took turns seizing power from each other. Plato sailed three times to Syracuse (as described in his long *Seventh Letter*), at the invitation of his friend Dion, in a doomed effort to help Dionysius the Younger improve his image by becoming a "philosopher king." The third visit must have been shortly before the end, in 357, of Dionysius the Younger's first period of rule. So the eclipse may have been that of 357 Feb. 29, though it would have been much less than total in Sicily..

331 Sep. 20, lunar total, just before the final battle of Gaugamela in which Alexander the Great of Macedon overthrew the last Persian emperor, Dareius III. Plutarch, *Life of Alexander* (xxxi 3): "It happened that in the month Boêdromiôn the Moon was eclipsed about the beginning of the Mysteries at Athens, and on the eleventh night after the eclipse, the armies being now in sight of one another, Dareius kept his forces under arms, and held a review of them by torchllight; but Alexander, while his Macedonians slept, himself passed the night in front of his tent with his seer Aristander . . . [He said:] 'I do not steal victory.'"

310 Aug. 5, solar total, across North Africa, Sicily, and Greece. Agathocles, tyrant of Syracuse in Sicily, was blockaded inside his city by the Carthaginians; he managed to break out, crossed the sea to land during an eclipse—presumably this one—in Africa and attack the Carthaginians in their homeland.

190 March 14, 129 Nov. 20, 127 Apr. 6, all **solar total**, are suggestions for the eclipse used by Hipparchus (about 190-120 B.C.) in calculating the distance of the Moon. They had similar north-trending tracks of totality, the first two across Greece and norrheastern Asia Minor, the third more southerly across Egypt, Arabia, and Persia.

168 June 21, lunar total. The Romans under Lucius Aemilius Paullus defeated the last Macedonian king, Perseus, at the Battle of Pydna, on the coast between Greece and Macedonia. Gaius Sulpicius Gallus, a scholar and close friend of Paullus, commanded the Second Legion: he predicted that there would be an eclipse of the Moon on the night before the battle. Paullus had him speak to the troops and explain this so that they would not be frightened. Nevertheless, Paullus was careful to make sacrifices and wait for favorable omens. The eclipsed Moon will have been low in the south. The Macedonians took it as an ill omen, foreboding the death of their king (though Perseus did not die till 166, in exile at Rome). Gallus on returning to Rome found that his prediction had brought him celebrity; he was elected consul, then played several other political roles, but continued his studies and in his later years turned mainly to astronomy.

A.D.:

33? During the crucifixion of Jesus, there was, according to the Gospels of Mark and Luke, darkness for three hours of the afternoon. But the execution happened at the Jewish Passover, which is at Full Moon. The gospels were probably written many decades later; various lunar eclipses have been suggested as sources of a confused memory; the nearest solar eclipse likely to be noticed was 29 A.D. Nov. 24, total in Damascus, about 97% at Jerusalem.

997 May 25, lunar partial. Two Persian mathematician-astronomers, Buzjani at Baghdad and Biruni at Bukhara (in what is now Uzbekistan), made coordinated observations which helped to determine the longitudes of the two places.

1030 Aug. 31, solar annular-total, across Norway and Sweden. Olaf Haraldsson, who had been baptized while visiting Normandy, became king of pagan Norway in 1016. Throughout his reign he fought ruthlessly against heathen practices, until the chieftains and farmers rose and drove him out. He returned in 1030 with a ragtag army, including his young half-brother Harald Hardradi. The Battle of Stiklestad was fought during this eclipse. Olaf was defeated and killed. Within a few hours, miracles were reported, and he became known as King Olaf the Saint. (Harald Hardradi later became king, invaded England in 1066, and was defeated and killed at the Battle of Stamford Bridge by Harold of England, just before Harold had to hurry south and be in his turn defeated and killed at the Battle of Hastings by Duke William of Normandy.)

1142 Aug. 22, solar total across Alaska, Canada, New York state. The founding of the Iroquois Confederacy (Haudenosaunee, Five Nations) is usually thought to have been in the 1400s or 1500s, but according to Barbara A. Mann and Jerry L. Fields it could have been on this date. The sky darkened in what apparently was near-total eclipse during the ratification council, which the Iroquois held in the afternoon at a place called Ganondagan, not far north of the path of totality.

1453 May 22, lunar partial. The Turks' siege of Constantinople began on April 6. Niccolò Barbaro, a Venetian physician, was in the city, and afterwards wrote an account. "At the first hour of the night, there appeared a wonderful sign in the sky, which was to tell Constantine [Constantine XI Palaeologus] . . . that his proud empire was to come to an end. . . . The Moon rose, being at this time at the full . . . but it rose as if it were no more than a three-day Moon, with only a little of it showing. . . . The Moon stayed in this form for about four hours." A week later, on May 29, the city fell, with massive effect on the history of Europe and the Middle East.

1485 Mar. 16, solar total across France. On that day died Anne Neville, one of the pawns in the Wars of the Roses. Daughter of Warwick the Kingmaker, she married Edward of Westminster and, after his death, Richard of Gloucester, becoming queen when he became King Richard III. She died probably of tuberculosis, but rumor had it that Richard had poisoned her and that the eclipse, though only partial in England, portended his fall, which happened two years later.

1504 Mar. 1, lunar total. Columbus, in his fourth and last voyage to the New World, had to stop on Jamaica in June 1503 because of the worm-eaten state of his ships. He sent a messenger to the Spanish governor on Hispaniola but had to wait a year for help. His hundred men ran out of food and depended on the Arawak natives, but they grew tired of the trinkets he offered in exchange. Like many navigators, Columbus carried a book of astronomical tables (it was the *Ephemerides* by Johann Müller of Königsberg, called Regiomontanus); in this he noticed the eclipse that was to start about sunset on February 29 (as seen from America). He called the Indians together and told them he was the servant of a god in the sky, who would cause the Moon to rise "inflamed with wrath." When the Moon rose with the typical reddish umbral shadow on it, the Indians came running with food. And the times of the eclipse contacts, or their difference from those given for Europe, enabled Columbus to calculate Jamaica's longitude; which should have told him he was nowhere near Cathay and the Ganges, as he still believed. (The Spaniards conquered Jamaica in 1509, under a licence from Columbus's son Diego, and soon exterminated the Arawaks.)

1560 Aug. 21, solar total. The path of totality crossed Africa and Spain. In Denmark the Sun was less than half covered, but the fact that it had been predicted confirmed the interest of Tycho Brahe, not yet 14, in astronomy. In France the prediction caused such panic that people fought to get into confessionals; a priest tried to calm them by announcing that the eclipse was delayed for two weeks.

1605 Sep. 27, lunar partial, and **Oct. 12, solar annular**. When Guy Fawkes nearly succeeded in blowing up the English parliament building on Nov. 5, it was obvious that this had been foreboded by the pair of eclipses recently seen, though for England the small lunar eclipse was around sunset and the solar was only partial, the central path passing southwest of Britain.

1715 May 3, solar total. "Halley's Eclipse": it crossed the middle of England, and he published a map of its predicted path, and afterwards a map corrected by about 20 miles. He and Newton observed it in London. They were joined by Jacques-Eugène d'Allonville, who had traveled from Paris and has therefore been called "the first eclipse-chaser." He may also have been doing some spying for France.

1806 June 16, solar total from California to New England. The Shawnee Prophet, Tenkskwatawa, brother of war leader Tecumseh, was challenged (it is said) by Governor Harrison of Indiana to show proof of his divine mission by, for instance, commanding the Sun to stand still or the Moon to alter its course. So the Prophet proclaimed that on this date he would darken the noonday Sun. The Indians, summoned to his village (at what is now Greenville, Ohio), watched it happen.

1831 Feb. 12, solar annular, seen across the southeastern U.S. Nat Turner in Virginia interpreted it as a black man's hand across the Sun, favorable sign for his slave uprising, which he planned to start on July 4 but postponed. A greenish Sun, or green halo around it, on Aug. 13 (not, as sometimes said, the next solar eclipse, total on Aug. 7 over the southern Pacific), prompted him to start on Aug. 21. An annular eclipse fails to cover the Sun, and Nat's rebellion was ruthlessly crushed.

1868 Aug. 18, solar total. Mongkut or Rama IV, king of Siam (Thailand), was a modernizer who made a study of Western science besides Buddhist astrology. Two years ahead, he calculated, even more accurately than French professionals, the path of totality across the southern isthmus shared by Thailand with Burma. He led a party of nearly 1000, including officials and foreign guests, with a fleet of steamers and 50 elephants, to a beach near Wakor. Clouds parted and the exactness of his prediction was a triumph for the king, but the sequel justified the foreboding of the astrologers. The camp, located as close as possible to the centerline, was on low ground infested by mosquitoes. Back in Bangkok, Mongkut died of malaria on his 64th birthday, Oct. 18. His 15-year-old son, Chulalongkorn, recovered from the disease, did not re-employ his tutor, Anna Leonowens, who was on leave in England, so she stayed there and wrote the memoir that became the musical *The King and I*.

Ian Dicks

1870 Dec. 22 solar total, along the Mediterranean. Pierre Jules Janssen, observing the 1868 Aug. 18 eclipse in India, made observations that suggested the element helium and the gaseous nature of the Sun's chromosphere; and in the same year Norman Lockyer of England discovered helium in the spectrum of the Sun and named it. Both these scientists led or took part in many eclipse expeditions. In 1870, Lockyer's went to Sicily; one of his supply ships was wrecked. Janssen escaped by balloon from the Prussian siege of Paris to join the French expedition at Oran in Algeria, but it was clouded out. Other expeditions went to Cadiz and Gibraltar.

1877 Aug. 23, lunar total. It is claimed that the Moon was seen partially eclipsed on the death night of Crazy Horse, captured war leader of the Oglala Lakota, but he was killed in the night of Sep. 5/6.

1878 July 29, solar total across the U.S. A small difference in Mercury's position from what was predicted in Newtonian theory was thought to be cause by a planet, "Vulcan," nearer than Mercury to the Sun. At this eclipse, Henry Draper, Thomas Edison and others tried to observe the planet; Lewis Swift and James Craig Watson believed that they did so.

1885. In Rider Haggard's novel *King Solomon's Mines* (published 1885) the English adventurers impress the Africans by predicting "an eclipse of the sun at the time of full moon." When the impossibility of this was pointed out, Haggard for the next edition changed the eclipse to a lunar one, which forced other changes in the story. And the impossible solar eclipse reappeared in some later editions.

1910 May 9, solar total across the Antarctic Ocean, partial in Australia. Later rumors were that this happened on the day before the death of King Edward VII of England was announced (though he died on May 6) and while Earth passed through the tail of Halley's Comet (though this was on May 19).

1919 May 29, solar total, South America to central Africa. Arthur Eddington was the English astronomer best placed to understand Einstein's 1915 general theory of relativity (because he was skilled in mathematics and free of anti-German prejudice). One of its predictions was that starlight passing close to the Sun would be bent by a certain small angle; this could be tested only during an eclipse. The 1914-18 war was still on, but an expedition was planned to observe the eclipse from Principe, then a Portuguese colony, now part of the two-island nation Sao Tomé and Principe, in the Gulf of Guinea. Eddington, a Quaker pacifist, was nearly conscripted for army service; the Astronomer Royal, Frank Dyson, won exemption for him on the ground of the expedition's importance. Eddington went to Principe, succeeded in photographing stars of the Hyades close to the eclipsed Sun, resulting in confirmation of Einstein and worldwide fame for him and Eddington.

1961 Feb. 15, solar total, low in the sky across Europe and western Asia. The makers of the film *Barabbas* shot this eclipse from central Italy to accompany in real time the scene of the crucifixion of Jesus. A logistical feat, but (see 33 A.D.) there could not have been a solar eclipse at that time.

1999 Aug. 11, solar total. Concorde 001, flying from London, followed the path of totality east across Europe so as to prolong the time of observation. The ends of the plane's great-circle route were about on the centerline; the middle of the route almost left the path's southern edge. On board were a dozen astronomers, with four large instruments aimed through special quartz windows in the roof and one through a side window.

Stories sometimes incorporate the misconception that the black Moon appears suddenly, and only in front of the Sun. From *Prisoners of the Sun* (adventures of Tintin) by Hergé; © 1949 by Casterman, Paris and Tournai. In Mark Twain's *A Connecticut Yankee in King Arthur's Court* (1889), engineer Hank Morgan, transported back to A.D. 528, similarly escapes burning at the stake by claiming to cause an eclipse.

The statistics of eclipses

Here is a census of the years 1901–2116:

				(yielding roughly:)	
				per century	per saros
all eclipses	980			454	82
eclipses of the Moon	491			227	41
penumbral		178		82	15
partial		135		63	11
total		178		82	15
eclipses of the Sun	489			226	41
partial		170		79	14
annular		158		73	13
including non-axial			5	2.3	0.4
no north limit			2	0.9	0.2
no south limit			1	0.5	0.1
annular-total		13		6	1
total		148		69	12
including non-axial			4	1.9	0.3
years with 4 eclipses	135			62.5	11
5 eclipses	54			25	4.5
6 eclipses	19			9	1.6
7 eclipses	8			3.7	0.7

The non-axial annular and total eclipses are those so close to the Earth's northern or southern rim (so close to being partial) that the antumbra or umbra touches but the axis of the shadow does not. The cases in our sample where the track of totality or annularity has no northern or southern edge are slightly less marginal, in that the axis does touch, though there could be other cases where it does not. Each of these species is rare enough that any given saros span is likely not to contain one; the whole span of time is not long enough to give a reliable statistic for them. Here are the dates of these borderline eclipses:

```
1918 May  19                              total non-axial
1950 Mar. 18          annular non-axial
1957 Apr. 30          annular non-axial
1957 Oct. 23                              total non-axial
1967 Nov.  2                              total non-axial
2003 May  31   annular no north limit
2014 Apr. 29          annular non-axial
2043 Apr.  9                              total non-axial
2043 Oct.  3          annular non-axial
2044 Feb. 28   annular no south limit
2101 Feb. 28   annular no north limit
2104 Dec. 17          annular non-axial
```

Annular-total eclipses, already borderline in another way, are unlikely also to be non-axial (the needle of the umbra falling just short and just wide of touching the northernmost or southernmost point of the Earth); they are said to achieve this only once in 250 million years.

"Central," or axial, solar eclipses embrace all the annular, annular-total, and total eclipses except the non-axial ones. They thus seem to be 310 in our sample; 144 per century; 26 per saros. What we might call "cordal" solar eclipses (involving the umbra or antumbra) would be 319 in the sample; 148 per century; 27 per saros.

The common and natural questions "How frequent are eclipses?," "How many are there per century?," "How many are solar and how many lunar, how many partial and how many total . . . ?" have answers that differ slightly depending on which span of time you are talking about. It's like asking how many cars are traveling along the highway: it depends which section you measure. This is presumably why an article in *Popular Astronomy* of Feb. 1976 said there are about 238 solar eclipses per century, of which about 84 are partial, 77 annular, 10 annular-total, and 66 total. These results must have been based on a different sample of years.

Finding the number per century (454) is useful in that it is virtually the same as finding the number per year (4.54), but the natural unit for counting off eclipses is neither century nor year: it is saros. *The frequency of eclipses is precisely the number of saros series that are running.* When the last eclipse of some series passes, as on 1902 April 8, the number of concurrent series decreases by one. Then when another series starts, as on 1917 July 19, the number increases by one. But series end and begin at very irregular intervals, as can be seen by searching for the symbols that mark them in our "Bead Curtain" chart. The best way to ask the statistical questions is "How many eclipses" (or "solar eclipses," etc.) "are there in a saros span of 18.03 years?" but then the better answer will not be from a single such span, but an average of many spans. That is why we have used our sample of 12 saros-spans, or two "saros centuries."

Another reason why the answers to these questions are variable is that calculations vary. Mine are not likely to be as accurate as those of Jean Meeus. In this span of years I found one partial solar eclipse where he found none (2051 May 10), one annular which he shows as partial (2086 Dec. 6), and 10 annular-total eclipses which he shows as annular (1927 Jan. 3, 1945 Jan. 14, 1948 May 9, 1963 Jan. 25, 1966 May 20, 1969 Mar. 18, 1984 May 30, 2002 June 10, 2020 June 21, 2085 Dec. 16; these are in a few saros series, especially series 137, which are in their stage of transition from one kind to the other). I don't feel so bad, because Oppolzer, the founder of the field, also got most of these things wrong the same way. I have altered my chart and table to fit Meeus's results. Based on my own calculations, there would have been 981 eclipses in the whole timespan, 490 solar, 149 of these annular and 23 annular-total.

With lunar eclipses, my results differ from those of Fred Espenak. He adds 6 penumbral eclipses where I found none (1951 Feb. 21, 1969 Aug. 27, 2016 Aug. 18, 2027 July 18, 2042 Oct. 28, 2096 June 6), turns two of my penumbrals into partials (1988 Mar. 3 and 2042 Sep. 29) and one partial into a total (2015 Apr. 4). This would raise the number of all eclipses and of lunar eclipses by 6, penumbral by 4, partial by 1, and total by 1, besides pushing the endings of lunar saros series 103, 108, 109, and 110 later, and the beginnings of series 151 and 156 earlier. However, Fred explains that his more generous number of eclipses, and promotion of slight to less slight types in borderline cases, is due to the figure he uses to represent the enlargement of the Earth's shadow because of dust in the atmosphere; and also that Meeus and the *Connaissance des Temps* did not recognize these eclipses. And, as he says, whether a doubtful penumbral eclipse occurs or not is of academic interest (which does not mean to say that it is of no interest) because it will be wholly unobservable. So I have decided to leave my lunar eclipses as they are.

Would our statistics become more refined if we took a census over many more centuries? Up to a point; but they would begin to become false. As with other astronomical averages such as the length of the year, there is ultimately no stable answer to such questions as "How frequent are eclipses?": over vastly longer time the answer changes. The Moon is receding from the Earth at a rate of 3.8 centimeters a year. In the far future it will appear too small to cover the Sun and there will be no total solar eclipses at all.

Census of the saros series

Let's return to the point that the frequency of eclipses, at a given time, is equivalent to the number of saros series running at that time. Here are the series running at some time during our 1901-2116 span, with the years when they begin and end.

	lunar series			solar series	
	begin	end		begin	end
102	479 —	*1958*			
103	472 —	*1933*			
108	689 —	*1951*	108	550 —	*1902*
109	736 —	*1998*			
110	747 —	*2009*			
111	830 —	*2092*	111	528 —	*1935*
112	859 —	2139			
113	888 —	2150			
114	971 —	2233	114	651 —	*1931*
115	1000 —	2280	115	662 —	*1942*
116	993 —	2291	116	727 —	*1971*
117	1094 —	2356	117	792 —	*2054*
118	1105 —	2403	118	803 —	*2083*
119	935 —	2396	119	850 —	*2112*
120	1000 —	2479	120	933 —	2195
121	1047 —	2508	121	944 —	2206
122	1022 —	2338	122	991 —	2235
123	1087 —	2367	123	1074 —	2318
124	1152 —	2450	124	1049 —	2347
125	1163 —	2443	125	1060 —	2358
126	1228 —	2472	126	1179 —	2459
127	1293 —	2555	127	991 —	2452
128	1304 —	2566	128	984 —	2282
129	1369 —	2613	129	1103 —	2528
130	1416 —	2678	130	1096 —	2394
131	1427 —	2707	131	1125 —	2369
132	1492 —	2754	132	1208 —	2470
133	1557 —	2819	133	1219 —	2499
134	1568 —	2830	134	1248 —	2510
135	1615 —	2877	135	1331 —	2593
136	1680 —	2960	136	1360 —	2622
137	1564 —	2953	137	1389 —	2633
138	1521 —	2982	138	1472 —	2716
139	1658 —	3065	139	1501 —	2763
140	1597 —	2968	140	1530 —	2774
141	1626 —	2888	141	1613 —	2857
142	1709 —	3007	142	1624 —	2904
143	1720 —	3000	143	1617 —	2897
144	1749 —	3011	144	1736 —	2980
145	1832 —	3094	145	1639 —	3009
146	1843 —	3123	146	1541 —	2893
147	1890 —	3134	147	1624 —	3049
148	*1973* —	3217	148	1653 —	2987
149	*1984* —	3246	149	1664 —	2926
150	*2013* —	3275	150	1729 —	2991
151	*2114* —	3358	151	1776 —	3056
			152	1805 —	3049
			153	1870 —	3114
			154	*1917* —	3179
			155	*1928* —	3190
156	2060 —	3503	156	*2011* —	3237
			157	*2058* —	3302
			158	*2069* —	3313
			164	*2098* —	3523

—95 series (47 lunar, 48 solar). During the span, 14 series end (6 lunar, 8 solar) and 11 begin (5 lunar, 6 solar). So 70 of the series (36 lunar, 34 solar) run throughout our span.

But what we really want to know is the cross-section across the series—the number running at one time. As the span opens, 84 are running (42 lunar, 42 solar). As each birth or death takes place, this number alters. As the span ends, 81 series are running (41 lunar, 40 solar).

		begin	end	after change:		
				lunar	solar	total
				42	42	84
solar	108		— 1902	"	41	83
solar	154	1917 —		"	42	84
solar	155	1928 —		"	43	85
solar	114		— 1931	"	42	84
lunar	103		— 1933	41	"	83
solar	111		— 1935	"	41	82
solar	115		— 1942	"	40	81
lunar	108		— 1951	40	"	80
lunar	102		— 1958	39	"	79
solar	116		— 1971	"	39	78
lunar	148	1973 —		40	"	79
lunar	149	1984 —		41	"	80
lunar	109		— 1998	40	"	79
lunar	110		— 2009	39	"	78
solar	156	2011 —		"	40	79
lunar	150	2013 —		40	"	80
solar	117		— 2054	"	39	79
solar	157	2058 —		"	40	80
lunar	156	2060 —		41	"	81
solar	158	2069 —		"	41	82
solar	118		— 2083	"	40	81
lunar	111		— 2092	40	"	80
solar	164	2098 —		"	41	81
solar	119		— 2112	"	40	80
lunar	151	2114 —		41	"	81

And so we see that the answer to the question "How many saros series are running?" varies (during this span) from 78 to 85 depending on the epoch (a word astronomers use to mean "the moment we are talking about").

At the epoch 1991, the number was 80 (41 lunar, 39 solar). It dropped by 1 with the quiet death of series 109 at the probably unnoticed penumbral lunar eclipse of 1998 Aug. 8. Even this was doubtful, since according to other calculations series 109 achieved yet one more and even slighter eclipse (skimming the penumbra for only 33 minutes) on 2016 Aug. 18. There was a question whether or not to include that in *Astronomical Calendaar 2016!*

Another lunar series ended on 2009 July 7, a solar started on 2011 July 1, and a lunar on 2013 May 25, so the total of concurrent series is back up to 80, where it will stay all the way to 2054 Aug. 3.

Eclipses to come

On the next pages are shown the paths of all the central solar eclipses (that is, annular and total) from 2017 to 2035. This span is chosen because, being 19 years, it is a Metonic cycle; which is a saros plus one year.

In it there are 28 of these more-than-partial solar eclipses: 12 total; 2 annular-total (2023 Apr. 20, 2031 Nov. 14); and 14 annular. One of the annular eclipses, 2020 June 21, is a borderline case that is annular-total according to some calculators. There are four years without any of these central solar eclipses (2018, 2022, 2025, 2029). During the same span, there are 14 partial solar eclipses, and 42 lunar, making 84 eclipses in all.

The eclipse paths are shown on five faces of the globe: the American, the Eurafrican, the Oceanian, and the north and south polar. **Boldface** labels and thicker lines indicate the total or annular-total eclipses. The globes are seen from a finite distance: 60 Earth-radii, the approximate mean distance of the Moon, so slightly less than a hemisphere is in view. Paths near the Earth's limb are compressed—some barely come into view—but they will be found on the neighboring pictures.

The collection of paths looks chaotic: all contrast, no pattern. It's like many lines of verse without rhyme. But that is because most of the eclipses fall within one saros, the span within which there is no repetition but which repeats as a whole. Because there is one year more, you can find, though not easily, the beginnings of rhyme: the eclipses of 2017 are mimicked, a third of the way westward around the world, by those of 2035. The annularity-path of 2017 Feb. 26 across the southern Atlantic reappears as that of 2035 March 9 across the southern Pacific, and the totality-path of 2017 Aug. 21 across the U.S. reappears as that of 2035 Sep. 2 across eastern Asia and the Pacific.

The eclipses stroke the globe at all manner of angles. But each is the angle appropriate to its time of year. Paths of the northern-summer months such as July arch northward because Earth's northern hemisphere is nodding toward the Sun; paths of November or January dip southward. Paths of northern spring slope up from south to north because Earth's north pole is then tipped backward (see the paths of 2024 Apr. 8 and 2023 Oct. 14, crossing in Texas). When looking at each path, you can see it as the line where a plane from the Moon cuts the Earth; this gives a sense of where the Moon is and which way the Earth is tilted.

Narrow paths are those of total and annular eclipses that verge toward each other: occasions when the umbra's tip barely reaches, or barely fails to reach, Earth's surface. They converge in the almost linear paths of the rare annular-total eclipses. Farthest apart in nature are the broadest path of totality (2027 Aug. 2, Egypt), when the umbra stubs deeply into the Earth, and the still broader paths of annularity when the Moon has withdrawn to its greatest distance.

The ellipses along the paths show the outline of the total or annular shadow, at 10-minute intervals. Even without labeling, the paths carry information about the times at which the eclipses occur, roughly. The middle position of the shadow is where the eclipse occurs at local noon.

For example, at the middle of the 2017 Aug. 21 path the shadow arrives at roughly longitude 90° west, which is where the Sun is at the meridian 6 hours after it reaches the meridian at Greenwich (longitude 0°). Therefore the time of greatest eclipse is about 6 hours after Greenwich noon: 18 Universal Time, or (naturally) 12 by local solar time. Counting the ten-minute shadows shows that the total eclipse begins and ends about 97 minutes before and after.

In these notes on the future eclipses, <u>underlining</u> draws attention to eclipses that are the subjects of whole sections elsewhere in the book.

<u>2017 Feb. 26</u> and <u>2017 Aug 21</u>: see the section on this year.

2018 is a year without central solar eclipses. (It has three partial solar and two total lunar eclipses.

2019 July 2, total: across the south Pacific; maximum duration 4 minutes 32 seconds (at northernmost latitude); coast of Chile (2m 30s); end at sunset just south of Buenos Aires. Saros 127 successor of 2001 June 21 across the south Atlantic and southern Africa.

2019 Dec. 26, annular: Qatar, Oman, south India, north Sri Lanka, Sumatra, Singapore, Borneo; Guam on centerline (duration of annularity 3m 10s). Saros 132 successor of 2001 Dec. 14, across the Pacific to Panama.

2020 June 21, annular: narrow path from sunrise on the River Congo to Eritrea, Yemen, Oman, Pakistan, India, Tibet, China, Taiwan; at its central point, in India near to a corner of Nepal, its duration drops to 38s; or, according to some calculations, it switches briefly to **total**. This is the most central member of saros series 137. Its predecessors were 2002 June 10, all annular, over the Pacific to just touch the coast of Mexico; and <u>1984 May 30</u>, the broken-ring eclipse across the eastern U.S.

2020 Dec. 14, total: south Pacific, southern Chile and Argentina (reaching 2m10s),

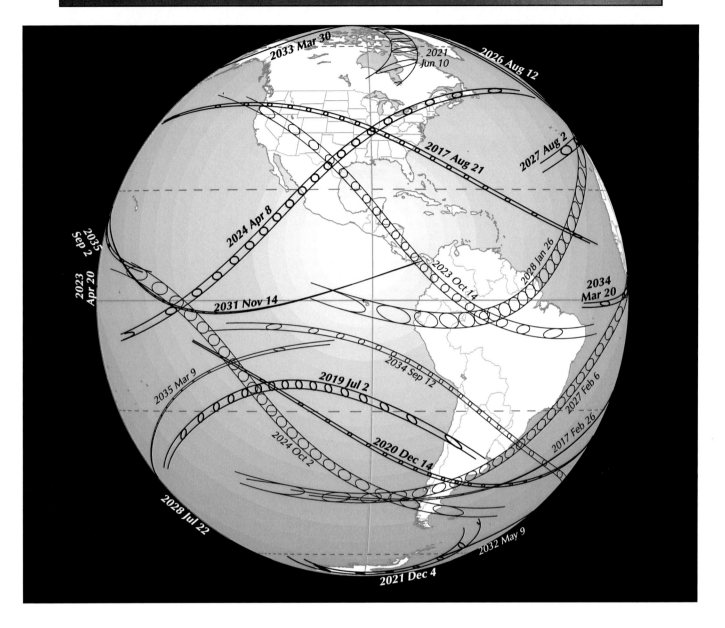

south Atlantic. Saros 142 successor of 2002 Dec. 4 over southern Africa, south Indian Ocean, Australia.

2021 June 10, annular: from eastern Canada via northern Greenland and Ellesmere Island to straddle the north pole and then south to eastern Siberia. Compare the 2015 March 20 total eclipse that ended almost at the north pole. Saros 147 successor of 2003 May 31, which made a semicircular track centered on Iceland, extremely wide because the "northern" edge was off the Earth.

2021 Dec. 4, total: across West Antarctica, that is, the peninsula south of South America; reaching 1m55s in the Weddell Sea. Saros 152 successor of 2003 Nov. 23, which made a semicircle from the Antarctic Ocean into Antarctica and back.

2022: no central solar eclipse. (Two partials, and two total lunar.)

2023 Apr. 20, annular at sunrise in the south Indian Ocean; the narrow path soon changes to **total**; just misses northwest corner of Australia; maximum duration 1m 16s just before touching northeast end of the island Timor; across the "head" of West Papua; changes back to annular shortly before ending at sunset almost exactly on the International Date Line (longitude 180°). Saros 129 successor of 2005 Apr. 8, also annular-total, over the south Pacific to Venezuela; and 1987 Mar. 29, Patagonia across the Atlantic to the Horn of Africa.

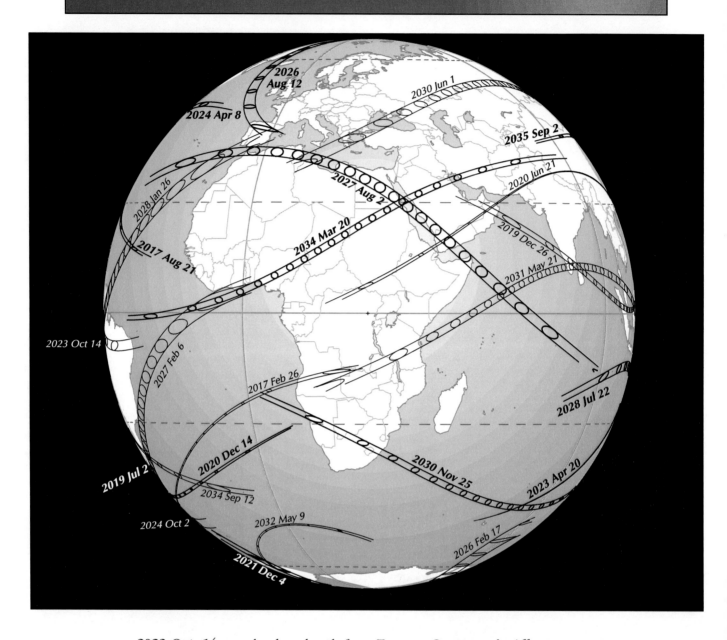

2023 Oct. 14, annular: broad path from **Eugene, Oregon, via Albuquerque to Corpus Christi, Texas**; then Yucatan and bits of all the countries of Central America (5m 17s in Nicaragua), Colombia, Brazil. Saros 134 successor of 2005 Oct. 3, from Spain to Libya and Kenya.

2024 Apr. 8, total: Pacific, Mexico (maximum 4m 28s near Durango), **Texas to Newfoundland**; in the path are Austin, Dallas, Indianapolis, Cleveland, Buffalo, Montreal. Crosses the 2017 Aug. 21 path in southern Illinois. Saros 139 successor of 2006 Mar. 29, from Nigeria to Libya-Egypt coast, Turkey, central Asia.

2024 Oct. 2, annular: central to south Pacific (reaching 7m25s), southern Chile and Patagonia. Saros 144 successor of 2006 Sep. 22, from the Guianas to the ocean south of South Africa.

2025: no central solar eclipse. (Two partial solar, two total lunar.)

2026 Feb. 17, annular: East Antarctica, i.e. the part south of Australia; reaching 2m 20s offshore in the Indian Ocean. Saros 121 successor 2008 Feb. 7, from the south Pacific into Antarctica.

2026 Aug. 12, total: starts near the north pole, barely missing it; reaches 2m18s just after leaving Greenland and before hitting Iceland; southward west of Ireland, to northern Spain and the Balearic Islands at sunset. Saros 126 successor of 2008 Aug. 1,

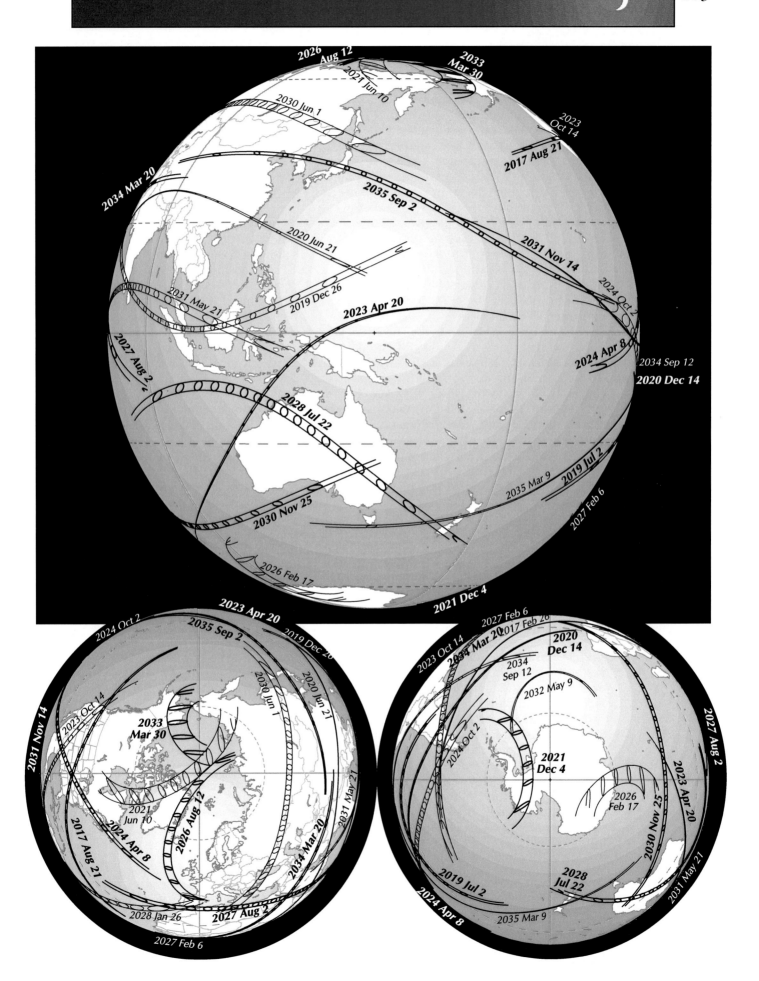

Canadian Arctic over north of Greenland to Siberia and China; and 1990 July 22, Scandinavia, north and east Siberia.

2027 Feb. 6, annular: south Pacific, Chile, Argentina, 7m51s near the coast of Uruguay; across Atlantic to sunset on the coast of Ghana. Saros 131 successor of 2009 Jan. 26, south Atlantic and Indian Oceans to Indonesia.

2027 Aug. 2. total: straddling the Strait of Gibraltar (southern Spain and Morocco); Algeria, Tunisia (Sfax near centerline), Libya (Benghazi on centerline); mid Egypt, Valley of the Kings just south of centerline, Luxor on it, maximum duration near there 6m 23s; Arabia (Jeddah, Mecca, and San'a in the path); tip of Somalia; Indian Ocean. Saros 136 successor of 2009 July 22 over China, and 1991 July 11 over Hawaii, Mexico, Central America.

2028 Jan. 26, annular: Galápagos Islands (almost all in the path); Ecuador, Peru, Brazil, French Guiana (reaching 10m 27s); sunset in southern Portugal, much of Spain. Saros 141 successor of 2010 Jan. 15: Congo to Kenya, south India, Burma, China.

2028 July 22, total: Indian Ocean; maximum 5m 10s on north coast of Western Australia; center of Sydney (3m 49s), South Island of New Zealand (centering Milford Sound and Dunedin, 2m 51s). Saros 146 successor of 2010 July 11: south Pacific to south end of South America.

2029: no central solar eclipse. (Three partial, two total lunar.)

2030 June 1, annular: Tunisia-Libya border, most of Greece, Istanbul, Crimea; along Russia-Kazakhstan border (5m 21s); Lake Baikal; Manchuria; north of Vladivostok; Hokkaido near sunset. Saros 128 successor of 2012 May 20, south China, Japan, Oregon-California to Texas; and 1994 May 10. across the U.S.

2030 Nov. 25, total: Namibia, Botswana, Lesotho, South Africa including Durban (2m35s on centerline); reaching 3m 44s in south Indian Ocean; south Australia, sunset west of Brisbane. Saros 133 successor of 2012 Nov. 13: north Australia, south Pacific.

2031 May 21, annular: Angola-Namibia border at sunrise; Zambia, and Lumumbashi in D.R. Congo; Tanzania including Dar es Salaam; Zanzibar on centerline (4m 16s); Kerala, Madurai, Jaffna in Sri Lanka; junction of Thailand and Malaya, Borneo, Sulawesi, at sunset in Maluku islands. This is the most central member of saros series 138. Its predecessor was 2013 May 10 (Australia, Papua New Guinea eastward).

2031 Nov. 14, annular in the Pacific, south of Hawaii, changing to **total** on the equator (duration up to 1m 8s), back to annular before ending in Panama. Saros 143 successor of 2013 Nov. 3 which was annular for only 15 seconds.

2032 May 9, annular: in the far southern Atlantic; duration at greatest eclipse only 22s. Saros 148 successor of the barely-annular 2014 Apr. 29 touching Antarctica.

2033 Mar. 30, total: broad oblique path from the Bering Sea and Strait (2m30s at Nome); reaches 2m 37s just after crossing Alaska; ends near north pole. Saros 120 successor of 2015 Mar. 20 that ended at the north pole.

2034 Mar. 20, total: mid Atlantic, Nigeria including Lagos; Chad (maximum 4m 9s), northwest Sudan, southeast Egypt; Arabia including Medina; southern Kuwait; Iran, Afghanistan, Pakistan including Islamabad, Tibet. Saros 130 successor of 2016 Mar. 9: Indonesia, north Pacific.

2034 Sep. 12, annular: Pacific, north Chile (2m 58s), Argentina, Uruguay, Atlantic. Saros 135 successor of 2016 Sep. 1: Atlantic, central Africa, Madagascar, Indian Ocean.

2035 Mar. 9, annular: off south of Tasmania; New Zealand South Island (including Westport, Nelson) and North Island (including Lower Hutt). Saros 140 successor of the eclipse of 2017 Feb. 26 over the south Atlantic.

2035 Sep. 2, total: Xinjiang north of Hotan at sunrise; Inner Mongolia; Beijing (4m50s in north of city); North Korea (Pyongyang 1m49s); Japan (Tokyo just outside south limit, 2m32s on centerline); Pacific near Wake Island. Saros 145 successor of 2017 Aug. 21 across the U.S.

This graph shows how solar eclipses compare in one respect: maximum duration of central eclipse. (Similar graphs could be drawn for centrality, nearness of Moon to Earth, or other qualities.)

All the central eclipses of the time-span are plotted, but only a few of the saros series are shown by connecting lines.

Open circles are annular eclipses; circles with dots in them, annular-total; solid circles, total. Annular eclipses can be much longer than total ones. The shortest maximum dura-

tions are for those near the border between annular and total. So the curve for a series that makes the transition from annular through annular-total to total bounces off the zero line.

But series 124 becomes partial after the annular-total eclipse of 1986. Series 146 passes from partial to total in 1938.

Series 136 peaked in the 20th century, with the eclipses of 1937, 1955, and 1973, all over 7 minutes.

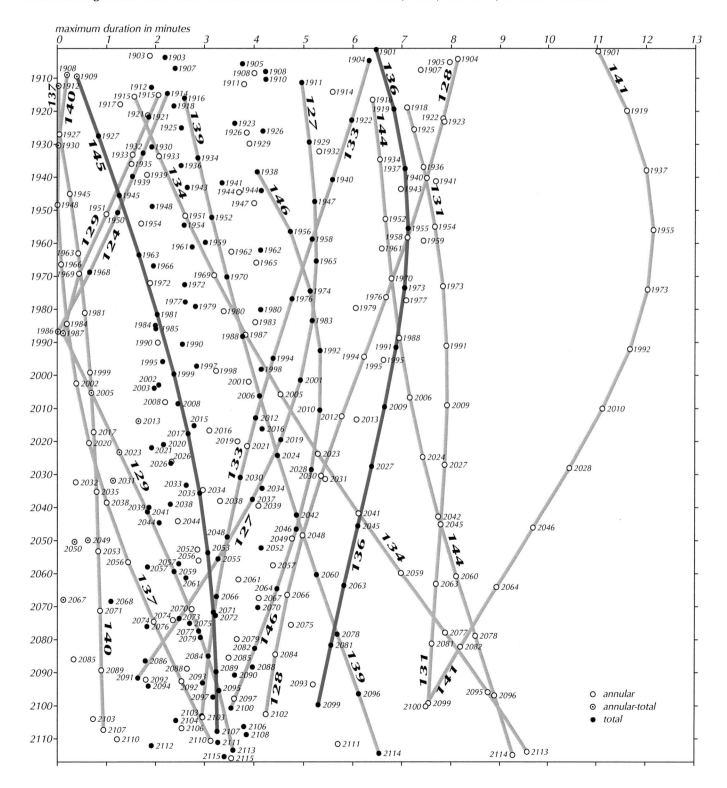

Tables of eclipses

Here is brief information about eclipses or sets of eclipses mentioned or pictured in the book.

"-674" means "675 B.C.," etc. This is the way calculators have to reckon with years before the beginning of an era. (There is no year "0 B.C."; so the year 1 B.C. is equivalent to 0, 2 B.C. to -1, etc.)

"h" stands for "hour," that is, the Universal (Greenwich) Time of mid eclipse. This is the time when the Moon is approximately at the meridian, so it also indicates which face of the Earth sees the eclipse. (For a lunar eclipse, middle longitude = (-hour) ×15. For a solar eclipse, middle longitude = (12-hour) ×15. If result is more than 180, subtract 360; if less than -180, add 360. Negative means west longitude.) Notice how in a saros series the hours roughly repeat in groups of three: "4 12 19 3 10 18 1 9 17" or the like.

You may wonder how the hour can sometimes be "24." The more exact time may be 23:47; if we were to approximate that to "0," we'd have to change the date to the next day.

In the "node" column, "Asc" and "desc" mean that the eclipse occurs at the ascending or descending node of the Moon's orbit.

"Gamma" is the axial distance: the distance (as a fraction of the Earth's radius) by which the axis of the Moon's shadow misses the central plane of the Earth; or, in a lunar eclipse, by which the center of the Moon misses the axis of the Earth's shadow. The nearer gamma is to zero, the more central the eclipse. When gamma is negative, the shadow passes to the south.

"Dist." is the distance between the centers of Moon and Earth, in Earth-radii.

The columns headed "m s" show the duration. That is, for solar total and annular eclipses, the maximum duration reached on the centerline. For lunar eclipses, it is the duration of the total phase.

In the saros columns, the first number is the saros series to which the eclipse belongs. The second is its order in that series. The third is its order relative to the central eclipse (0) of the series. "B" marks the first eclipse of a series, "E" the last.

Some eclipses of the ancient solar saros series 57

		h			node	gamma	dist.	m s	saros
-674 Apr.	5	6	SOLAR	TOTAL	Asc	0.658	56.02	4:57	57 28-14
-656 Apr.	15	14	SOLAR	TOTAL	Asc	0.596	56.02	5:11	57 29-13
-638 Apr.	26	21	SOLAR	TOTAL	Asc	0.530	56.03	5:25	57 30-12
-620 May	7	4	SOLAR	TOTAL	Asc	0.460	56.04	5:38	57 31-11
-602 May	18	12	SOLAR	TOTAL	Asc	0.388	56.07	5:50	57 32-10
-584 May	28	19	SOLAR	TOTAL	Asc	0.316	56.10	6:01	57 33 -9

Some eclipses used as examples

		h			node	gamma	dist.	m s	saros
1984 May	30	17	SOLAR	Annular	Asc	0.279	60.42	0:12	137 34 -1
1989 Mar.	7	18	SOLAR	partial	Asc	1.099	56.11		149 20-21
1989 Aug.	17	3	lunar	Total	Asc	-0.151	57.63	95	128 39 32
1986 Oct.	3	19	SOLAR	Ann-Tot	desc	0.995	58.66	0:01	124 53 20
1987 Mar.	29	13	SOLAR	Ann-Tot	Asc	-0.305	59.31	0:07	129 50 11
1990 July	22	3	SOLAR	TOTAL	desc	0.763	57.88	2:33	126 46 16
1991 July	11	19	SOLAR	TOTAL	desc	-0.002	56.09	6:53	136 36 0
1994 May	10	17	SOLAR	Annular	desc	0.409	63.58	6:13	128 57 13
2015 Mar.	20	10	SOLAR	TOTAL	desc	0.944	56.12	2:47	120 61 28
2017 Feb.	11	1	lunar	penumbr	Asc	-1.027	59.18		114 59 57
2017 Feb.	26	15	SOLAR	Annular	desc	-0.462	59.30	0:44	140 29 -4
2017 Aug.	7	18	lunar	Partial	desc	0.866	61.89	115	119 61 49
2017 Aug.	21	18	SOLAR	TOTAL	Asc	0.437	58.34	2:40	145 22-12

All eclipses of solar saros series 136

```
                 h                    node gamma dist.   m s    saros
1360 June 14  6 SOLAR  partial    desc-1.518 61.07           136  1-34B
1378 June 25 13 SOLAR  partial    desc-1.434 60.87           136  2-33
1396 July  5 19 SOLAR  partial    desc-1.352 60.66           136  3-32
1414 July 17  2 SOLAR  partial    desc-1.273 60.46           136  4-31
1432 July 27  9 SOLAR  partial    desc-1.197 60.25           136  5-30
1450 Aug.  7 17 SOLAR  partial    desc-1.125 60.04           136  6-29
1468 Aug. 17 24 SOLAR  partial    desc-1.059 59.84           136  7-28
1486 Aug. 29  7 SOLAR  (ann)      desc-1.000 59.63           136  8-27
1504 Sep.  8 15 SOLAR  Annular    desc-0.947 59.43  0:32     136  9-26
1522 Sep. 19 23 SOLAR  Annular    desc-0.900 59.24  0:23     136 10-25
1540 Sep. 30  7 SOLAR  Annular    desc-0.862 59.05  0:17     136 11-24
1558 Oct. 11 15 SOLAR  Ann-Tot    desc-0.830 58.86  0:12     136 12-23
1576 Oct. 21 23 SOLAR  Ann-Tot    desc-0.804 58.69  0:08     136 13-22
1594 Nov. 12  8 SOLAR  Ann-Tot    desc-0.785 58.51  0:04     136 14-21
1612 Nov. 22 16 SOLAR  Ann-Tot    desc-0.771 58.34  0:01     136 15-20
1630 Dec.  4  1 SOLAR  Ann-Tot    desc-0.760 58.18  0:07     136 16-19
1648 Dec. 14  9 SOLAR  Ann-Tot    desc-0.753 58.03  0:14     136 17-18
1666 Dec. 25 18 SOLAR  Ann-Tot    desc-0.748 57.88  0:24     136 18-17
1685 Jan.  5  3 SOLAR  Ann-Tot    desc-0.743 57.73  0:35     136 19-16
1703 Jan. 17 11 SOLAR  TOTAL      desc-0.737 57.59  0:50     136 20-15
1721 Jan. 27 20 SOLAR  TOTAL      desc-0.729 57.45  1:07     136 21-14
1739 Feb.  8  5 SOLAR  TOTAL      desc-0.718 57.32  1:27     136 22-13
1757 Feb. 18 13 SOLAR  TOTAL      desc-0.702 57.20  1:51     136 23-12
1775 Mar.  1 22 SOLAR  TOTAL      desc-0.681 57.08  2:19     136 24-11
1793 Mar. 12  6 SOLAR  TOTAL      desc-0.654 56.96  2:51     136 25-10
1811 Mar. 24 14 SOLAR  TOTAL      desc-0.621 56.85  3:26     136 26 -9
1829 Apr.  3 22 SOLAR  TOTAL      desc-0.581 56.74  4:04     136 27 -8
1847 Apr. 15  6 SOLAR  TOTAL      desc-0.535 56.64  4:43     136 28 -7
1865 Apr. 25 14 SOLAR  TOTAL      desc-0.483 56.55  5:22     136 29 -6
1883 May   6 22 SOLAR  TOTAL      desc-0.425 56.46  5:58     136 30 -5
1901 May  18  6 SOLAR  TOTAL      desc-0.362 56.38  6:28     136 31 -4
1919 May  29 13 SOLAR  TOTAL      desc-0.294 56.30  6:50     136 32 -3
1937 June  8 21 SOLAR  TOTAL      desc-0.224 56.24  7:04     136 33 -2
1955 June 20  4 SOLAR  TOTAL      desc-0.151 56.18  7:07     136 34 -1
1973 June 30 12 SOLAR  TOTAL      desc-0.076 56.13  7:03     136 35  0
1991 July 11 19 SOLAR  TOTAL      desc-0.002 56.09  6:53     136 36  0
2009 July 22  3 SOLAR  TOTAL      desc 0.072 56.06  6:39     136 37  1
2027 Aug.  2 10 SOLAR  TOTAL      desc 0.144 56.03  6:23     136 38  2
2045 Aug. 12 18 SOLAR  TOTAL      desc 0.213 56.02  6:06     136 39  3
2063 Aug. 24  1 SOLAR  TOTAL      desc 0.279 56.01  5:49     136 40  4
2081 Sep.  3  9 SOLAR  TOTAL      desc 0.340 56.02  5:33     136 41  5
2099 Sep. 14 17 SOLAR  TOTAL      desc 0.396 56.03  5:18     136 42  6
2117 Sep. 26  1 SOLAR  TOTAL      desc 0.446 56.05  5:03     136 43  7
2135 Oct.  7  9 SOLAR  TOTAL      desc 0.490 56.08  4:49     136 44  8
2153 Oct. 17 17 SOLAR  TOTAL      desc 0.527 56.11  4:36     136 45  9
2171 Oct. 29  1 SOLAR  TOTAL      desc 0.559 56.16  4:22     136 46 10
2189 Nov.  8 10 SOLAR  TOTAL      desc 0.584 56.21  4:09     136 47 11
2207 Nov. 20 18 SOLAR  TOTAL      desc 0.603 56.27  3:55     136 48 12
2225 Dec.  1  3 SOLAR  TOTAL      desc 0.618 56.33  3:42     136 49 13
2243 Dec. 12 12 SOLAR  TOTAL      desc 0.628 56.41  3:29     136 50 14
2261 Dec. 22 21 SOLAR  TOTAL      desc 0.635 56.49  3:16     136 51 15
2280 Jan.  3  5 SOLAR  TOTAL      desc 0.640 56.57  3:04     136 52 16
2298 Jan. 13 14 SOLAR  TOTAL      desc 0.645 56.67  2:52     136 53 17
2316 Jan. 25 23 SOLAR  TOTAL      desc 0.650 56.77  2:42     136 54 18
2334 Feb.  5  8 SOLAR  TOTAL      desc 0.657 56.88  2:33     136 55 19
2352 Feb. 16 16 SOLAR  TOTAL      desc 0.667 57.00  2:25     136 56 20
2370 Feb. 27  1 SOLAR  TOTAL      desc 0.682 57.13  2:17     136 57 21
2388 Mar.  9 10 SOLAR  TOTAL      desc 0.702 57.26  2:10     136 58 22
2406 Mar. 20 18 SOLAR  TOTAL      desc 0.728 57.40  2:03     136 59 23
2424 Mar. 31  2 SOLAR  TOTAL      desc 0.761 57.55  1:55     136 60 24
2442 Apr. 11 10 SOLAR  TOTAL      desc 0.800 57.71  1:45     136 61 25
2460 Apr. 21 18 SOLAR  TOTAL      desc 0.847 57.87  1:34     136 62 26
```

```
2478 May    3  2 SOLAR   TOTAL     desc 0.900 58.05  1:20 136 63 27
2496 May   13  9 SOLAR   TOTAL     desc 0.960 58.22  1:02 136 64 28
2514 May   25 17 SOLAR   partial   desc 1.025 58.41       136 65 29
2532 June   5  0 SOLAR   partial   desc 1.095 58.60       136 66 30
2550 June  16  8 SOLAR   partial   desc 1.170 58.79       136 67 31
2568 June  26 15 SOLAR   partial   desc 1.247 58.99       136 68 32
2586 July   7 22 SOLAR   partial   desc 1.327 59.19       136 69 33
2604 July  19  5 SOLAR   partial   desc 1.408 59.39       136 70 34
2622 July  30 12 SOLAR   partial   desc 1.488 59.60       136 71 35E
```

All eclipses from 2017 to 2035 (a Metonic cycle; a saros plus a year)

```
            h                  node gamma dist.  m s    saros
2017 Feb. 11  1 lunar   penumbr  Asc-1.027 59.18        114 59 57
     Feb. 26 15 SOLAR   Annular desc-0.462 59.30  0:44 140 29 -4
     Aug.  7 18 lunar   Partial desc 0.866 61.89115     119 61 49
     Aug. 21 18 SOLAR   TOTAL    Asc 0.437 58.34  2:40 145 22-12

2018 Jan. 31 14 lunar   Total    Asc-0.300 56.48 76     124 49 41
     Feb. 15 21 SOLAR   partial desc-1.214 62.49        150 17-21
     July 13  3 SOLAR   partial  Asc-1.352 56.05        117 69 33
     July 27 20 lunar   Total   desc 0.115 63.67103     129 37 32
     Aug. 11 10 SOLAR   partial  Asc 1.150 56.20        155  6-29

2019 Jan.  6  2 SOLAR   partial desc 1.138 63.13        122 59 25
     Jan. 21  5 lunar   Total    Asc 0.372 56.09 61     134 26 24
     July  2 19 SOLAR   TOTAL    Asc-0.644 57.65  4:32 127 58 16
     July 16 22 lunar   Partial desc-0.646 62.54177     139 21 16
     Dec. 26  5 SOLAR   Annular desc 0.411 60.25  3:40 132 46  8

2020 Jan. 10 19 lunar   penumbr  Asc 1.080 58.26        144 16  8
     June  5 19 lunar   penumbr desc 1.243 57.86        111 67 63
     June 21  7 SOLAR   Annular  Asc 0.124 60.83  0:39 137 36  0
     July  5  4 lunar   penumbr desc-1.368 59.44        149  3  0
     Nov. 30 10 lunar   penumbr  Asc-1.132 62.98        116 58 54
     Dec. 14 16 SOLAR   TOTAL   desc-0.296 57.13  2:10 142 23 -7

2021 May  26 11 lunar   Total   desc 0.477 56.05 15     121 55 46
     June 10 11 SOLAR   Annular  Asc 0.918 63.38  3:52 147 23-15
     Nov. 19  9 lunar   Partial  Asc-0.454 63.55207     126 45 37
     Dec.  4  8 SOLAR   TOTAL   desc-0.955 55.94  1:55 152 14-24

2022 Apr. 30 21 SOLAR   partial  Asc-1.188 62.18        119 66 30
     May  16  4 lunar   Total   desc-0.257 56.78 84     131 34 29
     Oct. 25 11 SOLAR   partial desc 1.071 59.01        124 55 22
     Nov.  8 11 lunar   Total    Asc 0.257 61.25 84     136 20 21

2023 Apr. 20  4 SOLAR   Ann-Tot  Asc-0.393 58.94  1:15 129 52 13
     May   5 17 lunar   penumbr desc-1.043 59.61        141 23 13
     Oct. 14 18 SOLAR   Annular desc 0.376 62.25  5:17 134 44  5
     Oct. 28 20 lunar   Partial  Asc 0.948 57.96 76     146 11  5

2024 Mar. 25  7 lunar   penumbr desc 1.063 63.56        113 64 59
     Apr.  8 18 SOLAR   TOTAL    Asc 0.345 56.41  4:28 139 30 -2
     Sep. 18  3 lunar   Partial  Asc-0.984 56.05 60     118 52 51
     Oct.  2 19 SOLAR   Annular desc-0.350 63.74  7:25 144 18-10

2025 Mar. 14  7 lunar   Total   desc 0.349 62.95 65     123 53 43
     Mar. 29 11 SOLAR   partial  Asc 1.043 56.24        149 22-19
     Sep.  7 18 lunar   Total    Asc-0.277 57.96 81     128 41 34
     Sep. 21 20 SOLAR   partial desc-1.062 62.26        154  7-27
```

```
2026 Feb. 17 12 SOLAR   Annular   Asc-0.974 60.28   2:20 121 61 27
     Mar.  3 12 lunar   Total     desc-0.377 59.99 57      133 27 26
     Aug. 12 18 SOLAR   TOTAL     desc 0.901 57.54   2:19 126 48 18
     Aug. 28  4 lunar   Partial   Asc 0.497 61.21197      138 29 18

2027 Feb.  6 16 SOLAR   Annular   Asc-0.297 63.14   7:52 131 52 10
     Feb. 20 23 lunar   penumbr   desc-1.051 56.96       143 18 10
     Aug.  2 10 SOLAR   TOTAL     desc 0.144 56.03   6:23 136 38  2
     Aug. 17  7 lunar   penumbr   Asc 1.281 63.51        148  4  1

2028 Jan. 12  4 lunar   Partial   desc 0.987 56.48 52      115 58 56
     Jan. 26 15 SOLAR   Annular   Asc 0.389 63.50 10:26 141 24 -5
     July  6 18 lunar   Partial   Asc-0.796 61.77139      120 58 48
     July 22  3 SOLAR   TOTAL     desc-0.604 57.10   5:09 146 28-13
     Dec. 31 17 lunar   Total     desc 0.329 59.20 70      125 49 39

2029 Jan. 14 17 SOLAR   partial   Asc 1.053 61.05        151 15-22
     June 12  4 SOLAR   partial   desc 1.294 61.72       118 69 32
     June 26  3 lunar   Total     Asc 0.009 58.52102      130 35 31
     July 11 16 SOLAR   partial   desc-1.418 60.10       156  2-30
     Dec.  5 15 SOLAR   partial   Asc-1.064 56.53        123 55 24
     Dec. 20 23 lunar   Total     desc-0.379 62.42 54      135 24 23

2030 June  1  6 SOLAR   Annular   desc 0.564 63.67   5:21 128 59 15
     June 15 19 lunar   Partial   Asc 0.754 56.25144      140 25 15
     Nov. 25  7 SOLAR   TOTAL     Asc-0.390 56.11   3:43 133 46  7
     Dec.  9 22 lunar   penumbr   desc-1.074 63.70       145 12  7

2031 May   7  4 lunar   penumbr   Asc-1.072 57.42       112 66 62
     May  21  7 SOLAR   Annular   desc-0.196 62.75   5:26 138 32  0
     June  5 12 lunar   penumbr   Asc 1.476 56.40       150  2 -1
     Oct. 30  8 lunar   penumbr   desc 1.185 60.29       117 53 53
     Nov. 14 21 SOLAR   Ann-Tot   Asc 0.306 58.25   1:08 143 26 -8

2032 Apr. 25 15 lunar   Total     Asc-0.354 60.57 66      122 57 45
     May   9 13 SOLAR   Annular   desc-0.937 59.70   0:22 148 22-17
     Oct. 18 19 lunar   Total     desc 0.421 57.19 45      127 42 36
     Nov.  3  5 SOLAR   partial   Asc 1.062 61.56        153 10-25

2033 Mar. 30 18 SOLAR   TOTAL     desc 0.977 56.08   2:38 120 62 29
     Apr. 14 19 lunar   Total     Asc 0.397 63.25 48      132 31 28
     Sep. 23 14 SOLAR   partial   Asc-1.158 63.61        125 57 20
     Oct.  8 11 lunar   Total     desc-0.289 55.95 78      137 27 20

2034 Mar. 20 10 SOLAR   TOTAL     desc 0.287 56.68   4:10 130 53 12
     Apr.  3 19 lunar   penumbr   Asc 1.119 63.36        142 19 12
     Sep. 12 16 SOLAR   Annular   Asc-0.395 61.42   2:58 135 40  4
     Sep. 28  3 lunar   Partial   desc-1.014 57.33 21      147  9  4

2035 Feb. 22  9 lunar   penumbr   Asc-1.038 59.01       114 60 58
     Mar.  9 23 SOLAR   Annular   desc-0.441 59.48   0:48 140 30 -3
     Aug. 19  1 lunar   Partial   desc 0.944 62.07 76      119 62 50
     Sep.  2  2 SOLAR   TOTAL     Asc 0.372 58.16   2:54 145 23-11
```

Some eclipse quantities

—some mentioned elsewhere; gathered here for convenience.

Width of paths of solar eclipses depends mainly on the width of the shadow as it reaches the Earth; which depends mainly on the distance of the Moon at the time. For a total-eclipse path passing near the center of the Earth (the ecliptic plane), width can be up to 269 kilometers (167 miles). For annular eclipse the path can be wider, because the Moon's average distance is greater than the length of its umbra (so that the antumbra can be broader at the Earth's surface than the umbra can). So the width can be up to 370 km (230 mi). Total-eclipse paths are widest at their middles, annular-eclipse paths toward their ends. Annular-total eclipses have the narrowest overall paths; width is zero at the points of change between annular and total. Paths crossing the northern and southern extremes of the Earth are greatly widened because the shadow strikes the surface obliquely.

Local frequency of solar eclipse. From the number of total eclipses (about 0.69 per year) and the area covered by each path (much harder to calculate), it is found that a point on the Earth is in the Moon's umbra once in 360 years—on average. The frequency is higher near the poles (because paths are wider), though this is offset by lowness in the sky and other inconveniences. It is also slightly higher in the northern hemisphere, because northern summer corresponds to the time when the Earth is on the outer half of its slightly non-circular orbit (aphelion, or farthest point from the Sun, is about July 4): traveling more slowly, it spends slightly more than half the year with more northern than southern hemisphere exposed to the Sun and therefore to solar eclipses. The Moon, too, being farther from the Sun throws an umbra that is longer and therefore broader where it reaches the Earth. This is said to change the averages as much as to one total solar eclipse in 320 years for points in the north. For particular places over limited times the frequency is likely to be far from average. This is clear in our globe pictures over a range of years: large areas untouched by eclipse paths contrast with small lozenge-shaped pieces of surface where paths cross, for instance the southern corner of Illinois, experiencing totality in 2017 and 2024. A sliver of inaccessible delta on the south coast of Papua New Guinea saw totalities less than a year and a half apart, 1983 June 11 and 1984 Nov. 22. There may be other spots deprived of totality for tens of centuries.

Speed of the shadow relative to the surface. The Moon moves across the Earth-Sun line at about 3400 kilometers / hour (2113 miles / hour); but the Earth is rotating in the same direction nearly half as fast at the equator (1670 km/hr or 1038 mi/hr), so the shadow's speed over the ground is about 1730 km/hr (1075 mi/hr) at its slowest, crossing the middle of the Earth. It is faster toward the polar regions (because they rotate more slowly), faster still beyond the poles (as at an Arctic eclipse in northern summer) where the surface is rotating in the opposite direction; and it approaches infinity as the shadow strikes the western (sunrise) side of the Earth or leaves the eastern (sunset) side, because here the surface points straight away.

Duration of eclipse at a spot follows from the speed of the shadow and the length across it at the spot. The peak duration usually quoted for total eclipses is reached at one spot on the centerline; it falls off for other points on the centerline, and for points nearer to the edge of the path, where it becomes zero. For totality, the longest possible is about 7 minutes 32 seconds, as of 2000, but will gradually decrease over the next 3,000 years to 7 minutes (see Jean Meeus, *Mathematical Astronomy Morsels III*, p. 60). The actual longest known of (2168 July 5) will be 7 minutes 26 seconds. By flying along the path, the duration can be lengthened: thus, on 1973 June 23 from 7 to 73 minutes! For annularity, the duration can be greater, since the width of the antumbra can be greater. The theoretical greatest is 12 minutes 30 seconds; the longest known (1955 Dec. 14) was 21 seconds short of this.

To find out more

The foundation work was Theodor von Oppolzer's *Canon der Finsternisse* [Canon of Eclipses] (1887), a catalogue and atlas of 8000 solar and 5200 umbral lunar eclipses from 1208 B.C. to A.D. 2161—all calculated by hand. (Previously there was only a sketchy eclipse catalogue by Pingré, 1766.)

There were several printed catalogues in the 20th century. They are now largely superseded by websites. Among many are these from NASA, covering all lunar and solar eclipse between 2000 B.C. and 3000 A.D.; for each total or annular solar eclipse, you can see maps down to detailed level and the circumstances for any point on Earth.
http://eclipse.gsfc.nasa.gov/LEcat5/LEcatalog.html
http://eclipse.gsfc.nasa.gov/SEcat5/SEcatalog.html

The *Astronomical Almanac*, a thick book of tables produced in Washington and London, and available in libraries, includes basic pages on the eclipses of the year ahead. A smaller version, *Astronomical Phenomena*, is published about a year earlier.

Fullest information about *The Total Solar Eclipse of 2017 August 21* is in a 146-page book with that title, by Fred Espenak and Jay Anderson, published in 2016.

Methods for calculating some of the many aspects of eclipses can, with much patience, be winkled out of the *Explanatory Supplement to the Astronomical Almanac* (new edition 1992). More aapproachable are Jean Meeus's *Astronomical Algorithms* (1999) and *Elements of Solar Eclipses 1951-2200* (1989), published by Willmann-Bell.

Magazines such as *Sky & Telescope*, *Astronomy*, and *SkyNews* (Canada) have varied articles about coming eclipses, especially their weather prospects, and act as clearing-houses for reports about eclipses afterward.

For the French-reading world, labor-of-love annual books are produced by Guillaume Cannat.

Index

For often-used terms only the first and a selection of other references are given. Not all places and dates mentioned are indexed. For some terms, short definitions are given in parentheses.

You go, and I must live for your return
The center of my cosmos is withdrawn
Dimmer and colder the forsaken land
There linger only shadows where you shone
Your wasting image shivers into sand
Now from the finger even the ring is gone
The only sun is embered to a Venus
What shimmers on is but a lovely ghost
Shall I remember this when you are back?
The alteration and the somber lack
If day can be denied what is to trust?
A globe it is that rolls its bulk between us
The world survives, a drained and frozen dance
Green is not green, life is but life suspended
Yet time is time and wakens from its trance
The reign of the unnatural is ended
You brush the globe aside, invent the dawn
Your rediscovered light will all the brighter burn

Made in the USA
San Bernardino, CA
12 September 2016